Achille Vintras

Medical Guide to the Mineral Waters of France and Its Wintering

Stations

Achille Vintras

Medical Guide to the Mineral Waters of France and Its Wintering Stations

ISBN/EAN: 9783337254605

Printed in Europe, USA, Canada, Australia, Japan

Cover: Foto ©berggeist007 / pixelio.de

More available books at **www.hansebooks.com**

MEDICAL GUIDE

TO THE

MINERAL WATERS OF FRANCE

AND ITS

WINTERING STATIONS

MEDICAL GUIDE

TO THE

MINERAL WATERS OF FRANCE

AND ITS

WINTERING STATIONS

BY

A. VINTRAS, M.D.

PHYSICIAN TO THE FRENCH EMBASSY
SENIOR PHYSICIAN TO THE FRENCH HOSPITAL IN LONDON
CHEVALIER DE LA LÉGION D'HONNEUR

LONDON
J. & A. CHURCHILL
11, NEW BURLINGTON STREET
1883

PREFACE

In offering to the Medical Profession—and through
them to the Public—this book on the Mineral Waters
of France, I have endeavoured to be as concise as
possible in order to furnish the busy practitioner with
the means of ascertaining at a glance the chemical
composition, the therapeutical value, and the geographical situation of the most important French
Thermal Stations. I have availed myself of the latest
monographs and other writings on the subject in order
to bring the information up to date; and if I am
unable to name all my French *confrères* to whom I am
under an obligation, it is partly because they are too
numerous, and also because I could not have named
them alone without doing some injustice to the hard-
working men of the past generation, who did so much
to increase the reputation of our mineral springs.

But there is one name which I could not omit to
mention—it is that of Dr Durand-Fardel, who has
perhaps done more than all others to elevate hydro-
therapeutics to the position of a special science. It is
almost impossible to open any book on mineral waters,

published within the last twenty years, without
recognising the thoughts—and often the very words—
of this eminent hydrologist. I am happy to acknow-
ledge here the benefits I have myself derived from his
works, and the heavy contribution under which I
have laid his admirable lectures on mineral waters,
in the Introduction of this book.

If I add that I have personally visited the most
important stations mentioned in this work, it will, I
hope, be taken as a guarantee of the genuineness of
the information supplied; and my labours will have
been amply repaid, if the result is to increase the
appreciation and the use of those means of healing
with which nature has so beneficently and abundantly
provided France.

June, 1883.

CONTENTS

INTRODUCTION

MINERAL WATER STATIONS

INTRODUCTION

THE STUDY OF MINERAL WATERS

THE use of mineral waters occupies a place of no slight importance in the treatment of chronic diseases. There are few maladies of long duration in which they cannot be beneficially used at some period of their course ; there are few conditions of the constitution which they cannot advantageously modify. They place at the disposal of the physician powerful actions, from which effects may be obtained that would be expected in vain from any other therapeutical agent. With their assistance, and by availing himself of their diversity of composition, variety of thermality, and numerous methods of application, he can obtain substitutive, restorative, alterative, sedative, and resolvent actions.

That which chiefly distinguishes mineral waters from pharmaceutical remedies is that they are the best representatives of those general medications which exercise a very extended action upon the constitution, on the whole of the secreting system, on the capillary

1

circulation, and on the ultimate phenomenon of nutrition. Consequently, in many cases, a result can easily be obtained by the use of mineral waters, which all the resources of ordinary treatment would fail to achieve.

The principal aim in studying the subject of mineral waters is to learn how to make use of them, as of other therapeutical agents; and this knowledge has a very special character. A mineral water is not a medicine which can be kept at hand and used at will, or experimented with and tested by an immediate application. It is a medication at a distance, which he who prescribes it can neither test nor use himself; the indications which call for it, admit of no hesitation, yet the decision is complicated by questions of climate, distance, length of absence from home,—a number of conditions, in fact, which are either absolutely imperative, or of which account must at least be taken.

Among the knowledge necessary to the physician, an acquaintance with the subject of mineral waters seems, therefore, of high importance. Clinical study in hospitals and in private practice teaches nothing on this point, and individual experience is only obtained by special investigation.

The number of mineral springs existing on the surface of the globe is immense; it would even be difficult to make an exact enumeration of those in France alone. But it is evident that the physician needs only to be acquainted with those of which he may have occasion to avail himself, i.e. those which either possess suitable local arrangements for the treatment of patients, or whose waters can be used at a distance.

It ought to be known that—just as medical treatment can be carried on with a small number of drugs, so thermal treatment can be carried on with a small number of mineral springs. Nevertheless, it would be wrong and unwise, in thermal as well as in ordinary therapeutics, to neglect the varied agents which make it possible to comply with the indications of infinitely varied symptoms, and even with the preferences of the patient.

It is possible, therefore, while retaining a very wide range of choice in every division of treatment, to limit the number of Thermal Stations, the knowledge of which is important to every physician, by basing the selection upon the following considerations. Thermal medication cannot be brought to the patient; he must go to it, and his medical adviser alone can judge of the reasons which render it desirable that he should seek it near home, or at a distance. Questions of climate and altitude have to be considered as well as those purely medical; and there are often personal reasons for selecting an animated, brilliant, and luxurious Station, or one where a residence will be simple and quiet.

We shall, therefore, confine ourselves in this work to the consideration of those Thermal Stations which, by their therapeutical value, their reputation, the present state of their arrangements, and their geographical or climatic conditions, furnish the elements of a complete thermal treatment. We shall find these requirements almost entirely fulfilled by the mineral springs of France.

Before studying the therapeutical use of mineral waters, it is necessary to be acquainted with their actual nature, their geographical distribution, and their special modes of application.

DEFINITION OF MINERAL WATERS

How can mineral waters be defined? It cannot be done by enumerating the characteristics proper to their composition, for it is in this respect that they differ most widely from each other. Nor can the proportion of the mineral principles which they contain be taken as distinctive, since some of the most interesting are the least strongly mineralised; nor is temperature a test, since some are cold, and some thermal, or hot.

The only clear definition is supplied by the use which is made of them; and the term *mineral waters*, when used in medicine, may be understood to mean *natural waters which are employed in therapeutics, in consequence of their chemical composition or temperature.*

ORIGIN OF MINERAL WATERS

Natural mineral waters consist of water and mineralising principles. The water comes from the atmosphere, the principal constituents of which (oxygen, nitrogen, and carbonic acid) are to be found in the springs. The mineralising principles come from the soil, from the primitive rocks, or the sedimentary strata.

In respect of their origin, mineral waters are divided naturally into two groups : (1) those called the deep, geological, or overflowing, waters ; and (2) the superficial waters of infiltration, or of lixiviation. The first are forced out of their subterranean reservoir by the incessant influx of atmospheric waters, and probably owe the force with which they ascend to the surface, rather to the expansion of gas and vapours than to the operation of the law—that water finds its own level. Properly speaking, they represent the emanations of aqueous volcanoes, the product of a real sublimation. They are almost all of high thermality, strongly mineralised, and generally abundant in quantity. The waters of the second group result from the infiltration of rain water through the sedimentary strata, either upper or middle. In this case, pressure and elevation play a less important part; and the essential agent in the mineralisation is the carbonic acid, which the atmospheric waters bring with them, or meet in the soil. They return to the surface, like the waters of artesian wells, in order to find their level. They have usually a low thermality (*i.e.* are cool, or cold), are only feebly mineralised, and are less abundant than those of the other division.

There are also mineral waters that may be called *intermediary*, obtained by means of artesian borings, and proceeding either from ascending or descending waters, which have been stopped in their progress by meeting with impermeable strata. These have a certain artificial character, which clearly distinguishes them from those springs that issue spontaneously from the earth.

TEMPERATURE OF MINERAL WATERS

Mineral springs present the most varied temperatures, ranging from 10° C. (50° F.) up to 100° C. (212° F.) They may be classified according to their temperature, as follows :

Cold   .   Below 18° C. (64° F.)
Warm   .   From 18° to 28° C. (64° to 82° F.)
Hot   .   From 28° to 36° C. (82° to 97°. F.)
Very hot .   Above 36° C. (97° F.)

These designations are of consequence, in connection with the advantage that can be taken of their differing temperatures, and the action which may be attributed to these, either in the internal or external use of the waters.

There appears to be no doubt that the thermality (or natural heat) of mineral springs is principally due to the inner temperature of the earth at the point where they are collected; but this does not imply that the chemical reactions which there take place have no influence upon it. Different springs at the same Thermal Station have always different temperatures.

The thermality of mineral springs has no specific value. It may be considered that the higher the thermality, the more directly the water has risen from its bed, and the more closely it has preserved its original composition. It is, however, certain that too high a temperature forms rather an embarrassment than an advantage in the use of the waters; and that the most favorable thermalities are those which are best adapted for bathing purposes, and approach most nearly to the temperature of the blood, *i.e.* from 28° to 36° C. (82° to 97° F.)

CHEMICAL CONSTITUTION OF MINERAL WATERS

It must not be forgotten, in reading an analysis of a mineral water, that the combinations in which its constituent elements are presented are, to a certain extent hypothetic. The methods of analytical procedure do not permit the extraction of these bodies in the form of salts, but merely as acids and bases; and it is only in virtue of synthetic principles that these acids and bases can -be brought into connection with each other, and expressed in definite proportions. But these principles are not absolute, and different chemists may apply them differently.

Nevertheless, as the separate table of acids and bases alone does not convey sufficient meaning to physicians, and as after all the dissidence of the chemists cannot bring any serious modification to the therapeutical character of a mineral water,—we must still have recourse to the hypothetic analysis, to which it is now usual to append the actual analysis.

At the outset, it would appear as if the multiplicity of the substances which enter into the composition of the greater part of mineral waters, must render the correct understanding of their proper or relative compositions a very difficult matter.   But in reality it is not so, and the apparent difficulty can be greatly reduced.

The following table gives the principal elements in the composition of mineral waters, and will serve to show their manifest importance and significance.

Table of Substances most usually met with in Mineral Waters, and which ought almost exclusively to be considered.

ACIDS.	BASES.
Carbonic	Soda
Sulphuric	Potash
Hydrosulphuric	Lithia
Hydrochloric	Lime
Hydroiodic	Magnesia
Hydrobromic	Manganese
Arsenic	Iron
Silicic	Copper

GAS.

Carbonic
Sulphuretted hydrogen
Nitrogen

Nitrogeneous organic matter

This table includes only the enumeration of the substances which analysis has most generally discovered in mineral waters, and to which a distinct action may be attributed, either chemical or therapeutical. There are others which may be neglected, either because of their rare occurrence, or want of distinct significance. This does not imply that they serve no purpose in the waters in which they are found, but that they play either an apparently subordinate part, or one to which it is actually impossible to assign its place in medical hydrology.

If the *oxygen* introduced by the air might be neglected, except for the part that it takes in the chemical transformations of certain mineral waters, the *nitrogen*, on the other hand, deserves notice, if only on account of its abundance, although no distinct therapeutical effect can as yet be attributed to it.

Carbonic acid is, with the acids of sulphur (and leaving aside the chlorides), the essential agent in the mineralisation of the springs. Most of the bases of the mineral waters owe their soluble condition to its presence in excess. It exists equally in the free state, and there are districts (like those of Vichy and Vals in France, and Marienbad in Bohemia) where the ground is literally infiltrated with it. Pressure appears an important element in the combinations into which it enters, for, as soon as that is removed, the carbonic acid manifests a tendency to disengage itself and become isolated. Its presence is an almost essential condition of the internal use of all mineral waters, except the sulphureous group.

Iodine is nearly always in the imponderable state when found in mineral waters, and it is noteworthy that it has always a tendency to combine with the organic matters in the water, just as the iodine of the sea can seldom be detected except in sea-weeds.

Bromine exists in much larger proportions, but almost exclusively in chlorinated waters.

Silica, which—with the *humus*—forms the base of the soil, and is almost the sole constituent of the primitive rocks, is to be met with in almost all mineral waters, in an ill-defined state, either as base or as acid, but it is not possible to attach to it any precise therapeutical significance.

Iron is also present in almost all mineral waters, where it is held by carbonic acid, and sometimes, though rarely, by sulphuric acid; but it is not always found in sufficient quantity to produce sensible therapeutical effects.

Arsenic has been long recognised among the components of mineral waters. It habitually accompanies iron, especially in the mountainous region of Auvergne; but is also sometimes found in combination with soda, and perhaps potash. In a small number of mineral waters it attains a distinctly therapeutical proportion, but even then a preponderating position can seldom be assigned to it.

Soda exists everywhere;—not to speak of the salt of the sea,—it is present in the soil, in animal organisms, in our food, even in the dust which we inhale. It is found in the greater number of mineral waters, usually in predominating proportions; and its importance there is such that the therapeutical value of a mineral water can be almost accurately measured by the quantity of soda which it contains.

Potash scarcely appears in mineral waters, and then does not seem to play any distinguishable part in their action.

Lithium ought also to be mentioned, its presence having often been detected by spectrum analysis.

Lime and *Magnesia*, which are earthy bases, hold (especially the first) too large a place in the soil not to be carried away by almost every mineral spring, but the predominance of either confers on a mineral water a therapeutical value far inferior to that bestowed by the predominance of soda.

It is not necessary to do more than mention the *crenic* and *apocrenic* organic acids, which exist in mineral waters, but whose constitution is still ill-defined. According to some analyses, they are supposed to be in combination with iron.

All mineral waters, like soft waters, contain organic matters in a state of dissolution; but a large number are peculiarly rich in products of this class, which are distinguished as *organised* and *organic* matters.

Organised matter is composed of an amorphous substance, without consistency, of various colours, appearing only under the influences of air and light (above all, of solar light), and showing signs of organisation in cells or filaments. This is what is called *barégine* or *glairine*.

Organic matter is in a more advanced stage of organisation, and is represented in mineral waters by confervæ of the *phycean* class, which attract and hold the iodine of the waters, as the *fucus* does the iodine of the sea. They have been called *sulfuraires*, because they are formed chiefly in sulphureous waters, and sometimes in extraordinary proportions. The waters of Amélie-les-Bains, for instance, would give more than 1500 pounds per day of dry organic matter ; and other springs in the same neighbourhood would yield even more.

What is the origin of these organic matters, and of the multitude of microscopic infusoria, which are also to be met with in mineral waters ? Are they formed in these mineral waters themselves, or brought up from the earth, or due to germs conveyed by the atmospheric waters ? This is what has not yet been determined. These organic matters cannot certainly be credited with any share in the therapeutical action of the waters, but they impart to some springs a peculiar softness, which is taken advantage of in the bathing treatment.

CLASSIFICATION OF MINERAL WATERS

The classification of mineral waters is based on the predominance of one or several mineralising principles.

Those which can be taken into consideration in the classification are enumerated in the following table :

Substances which are alone useful in *Classification.

ACIDS.	BASES.
Carbonic	Soda
Sulphuric	Lime
Hydrosulphuric	Magnesia
Hydrochloric	Iron

It is to be noted that the latter (and they are not the only substances of which the same remark may be made) are to be found in almost all mineral waters, though in differing proportions. Thus, out of 97 springs,

Chlorides have been found in	. . . .	95
Carbonates ,, ,,	. . .	93
Sulphates ,, ,,	. . . .	89
Lime has been found in	. . . .	95
Soda ,, ,,	. . . . .	92
Magnesia ,, ,,	. . . . .	90

The reason of this is that these identical principles are contributed by an identical medium—the earth. In the same manner, we find that in the vegetable kingdom distinct alimentary or therapeutical principles are surrounded by the same chemical or organic elements, which are common to all its products.

A mineral water ought to be considered as a whole, from which no part can be abstracted without loss of identity. Nevertheless, in each particular water certain principles predominate, and characterise it both hydrologically and therapeutically.

The distinguished Professor Durand-Fardel has long ago pointed out that mineral waters which are grouped together by the community of one predominating principle, possess at the same time identical therapeutical properties ; and this he has called the *specialisation* of mineral waters.

Consequently, all practical study of mineral waters must start from their *classification* and *specialisation*, and then the chemical classification becomes really a therapeutical classification.

All such classification is based upon the predominance of a chemical principle, that is, of a salt ; and as the chemical and therapeutical characteristic of a salt is chiefly derived from its acid, the divisions of mineral waters are fixed by the consideration of the acids.

In this manner, the following classes have been formed :—*Sulphuretted, chlorinated, bicarbonated,* and *sulphated.*

The relative predominance of the bases serves to establish divisions within the classes. These predominant bases are always either soda or lime, and occasionally magnesia, which, however, usually accompanies lime.

There are, therefore, *sodic* waters, and *calcareous* or *magnesian* waters,—two groups which were formerly called *alkaline* waters, and waters with *earthy bases.* But as there are also others in which it is impossible to recognise a formal predominance of either, an additional group must be formed, consisting of *mixed waters, i.e.* in which the sodic and earthy bases are equally proportioned.

The same is the case with the acids. There are certain mineral waters which present several predominating acids; and thus chlorides are found with sulphates, and bicarbonates with chlorides. It is necessary, therefore, to form distinct classes for these, reserving the name of *groups* for the leading types presented by the old classes—sulphuretted, chlorinated, bicarbonated, and sulphated.

The number of *classes* defined is notably increased by this nomenclature; but it adds to the lucidity of the therapeutical differentiation, while leaving unaltered the essential divisions constituted by the old classes, to which is given the name of *groups.*

It must not be supposed that predominance is decided by the mere figures furnished by the chemical analysis. Therapeutical predominance has also to be taken into account. Such is the case with *sulphuretted waters,* in which the sulphurets exist in an inferior proportion to the other substances, but yet impart to the waters of this group a chemical and therapeutical character which is most distinct.

In the same manner, we recognise a class of *ferruginous waters,* although the iron is always present in a secondary proportion, and (what is still more important) only in the condition of a base. This is the reason why the ferruginous waters form only a *supplementary* group.

Lastly, there are waters so feebly mineralised that they offer in reality no predominating principle, and can be included in no class. The Germans have called these *indifferent* waters; but this is an improper designation, since they are far from being indifferent in their applications. It seems preferable to form a special group under the name of *indeterminate* waters, a word which expresses a double truth, both chemically and therapeutically, since it is impossible to infer from their composition any determinate application.

This group of the Indeterminates includes two classes:

(1) *Simple thermal* waters, which essentially represent it.

(2) *Feebly mineralised* waters, which possess special properties, sufficiently marked to distinguish them from the preceding class, yet not sufficiently so to attach them to any other family.

Such are the principles that have guided Dr Durand-Fardel in his new *Classification*, of which the following is a tabular view:

CLASSIFICATION OF MINERAL WATERS

GROUP :—Sulphuretted.

(One Class.)

CLASS I. Sulphuretted.—*Division* 1. Sodio-sulphuretted.
                        2. Calcareo-sulphuretted.

GROUP :—Chlorinated.

(Four Classes.)

CLASS   I. Sodio-chlorinated.
        II. Chloro-sulphuretted.
        III. Chloro-bicarbonated.
        IV. Chloro-sulphated.

GROUP :—Bicarbonated.

(Four Classes.)

CLASS   I. Bicarbonated.—*Division* 1. Sodio-bicarbonated.
                               2. Calcareo-bicarbonated.
                               3. Mixed bicarbonated
                                   (sodio-calcareous.)
        II. Bicarbonato-chlorinated.
        III. Bicarbonato-sulphated.
        IV. Bicarbonated, mixed (sulphates and chlorides).

GROUP :—Sulphated.

(One Class.)

CLASS I. Sulphated.—*Division* 1. Sodio-sulphated.
                       2. Calcareo-sulphated.
                       3. Mixed sulphated (sodio calcareous).
                       4. Magnesio-sulphated.

GROUP :—Indeterminate.

(Two Classes.)

CLASS   I. Simple thermal waters.
        II. Feebly mineralised waters.

SUPPLEMENTARY GROUP:—Ferruginous Waters.

The study and classification of mineral waters, such as has now been laid before the reader, leads to a very important consideration. It is, that according as we descend from the first group upon the list (that of the *sulphuretted* waters) to the last (the *indeterminate*), we find their characteristics becoming less marked, and their therapeutical effect growing feebler. Similarly, in each class, as we pass from the *sodic* divisions to the *calcareous* or *mixed* divisions, we find their application subject to less distinct indications, and producing less energetic actions.

The class of ferruginous waters is called *supplementary*, because it stands outside the classes which have been based upon the predominance of acids, and for this reason. Iron exists in almost all mineral waters, but only a certain number contain it in quantity sufficient to be taken into account. Even among these, the ferruginous quality is often overshadowed by other more essential principles, which more directly govern their specialisation.

On the other hand, a mineral water which contained *only* iron would be simply a ferruginous medicine like any other, better perhaps, but without those qualities which it is impossible to reproduce artificially, and which natural mineral waters derive from their complex composition.

We should therefore conclude that no mineral waters ought to be admitted into the ferruginous class, except those in which iron exists in therapeutical proportions, while the other principles are not present in sufficient quantities to impart to them a different character.

Iron is always found as a base; that is, generally as a bicarbonate, rarely as a sulphate, more often as an arseniate or combined with manganese. It is also associated with crenic or apocrenic acids, substances still very ill defined.

Although iron is itself present in but small proportion, yet it is certain that the form under which it exists, and its association with the principles which are found accompanying it in mineral waters, impart to it a therapeutical activity far exceeding that of all the ferruginous preparations contained in the Pharmacopœia.

PHYSIOLOGICAL AND THERAPEUTICAL ACTIONS OF MINERAL WATERS

The study of the various classes of mineral waters cannot be separated from the examination of the properties attributed to them, for the range of their medicinal applications is very extended. Entirely unsuited to acute complaints, there are few chronic diseases in which they cannot be used, with at least some advantage. Mineral waters can be applied (1) to definite diathetic diseases; (2) to less clearly-defined states of the constitution, which may even barely fall within the limits of pathology; (3) to diseases of the organs or functional mechanism which are in general connected more or less directly with constitutional or diathetic conditions, or else with certain conditions of the abdominal venous circulation which are called *plethora*, or *abdominal venosity.*

When thermal treatment is applied in a suitable and methodic manner, the following effects are observed : appetite is developed, digestion is effected with greater ease, the skin performs its functions with increased activity, the blood circulates more freely, the glandular secretions are stimulated, there is a tendency to the development of hæmorrhoidal phenomena—or else the catamenia appear or become more abundant, the temperature of the body is raised, and strength is augmented.

All these symptoms—which follow from the general stimulation of the functions, and lead to a definitive reconstitution of the organism—are the expression of the action of the thermal treatment upon the healthy portion of the system. But mineral waters also act directly upon unhealthy organs ; and by the penetration of the elements which they contain into the innermost recesses of the circulation and nutrition, they assist directly in the resolution of obstructions and tumours. Thus, also, by stimulating in a special manner the intestinal, renal, and cutaneous secretions, they produce revulsions by purgative, diuretic, or sudorific action.

The great characteristic of mineral waters is that they represent an alterative medication, *i.e.* a medication endowed with those essential properties by virtue of which the treatment changes the mode of being of the organism, by affecting the innermost phenomena of the nutrition. All these actions—whether alterative, reconstitutive, substitutive, or revulsive—lead to special actions, which are divided among the different groups, according to the analogy of their constitutions.

It is the latter that constitute the *specialisation* of the action of mineral waters, a specialisation which (besides the attributes proper to each individual spring) marks with a significative character each one of the classes which have been established.

Just as the knowledge of mineral waters is entirely based upon their classification, so the knowledge of their application rests entirely upon their specialisation. The latter, in a manner, furnishes the proof of the legitimacy of the classification, by showing that it possesses the characters of a natural and practical method.

The following table reproduces the formula of these *specialisations;* but it must not be forgotten that however exactly they may be stated, it is necessary to penetrate further into the study of mineral waters before all their applications can be rightly grasped.

TABLE OF THE SPECIALISATION OF MINERAL WATERS

GROUP:—Sulphuretted.

Divisions :—Sodio- or calcareo-sulphuretted.

Special applications:—Herpetism, dermatoses, catarrh of the respiratory passages.

General applications :—Lymphatism, rheumatism, chlorosis, syphilis, scrofula.

Secondary applications:—Surgical complaints, chronic metritis, catarrh of the urinary organs, dyspepsia.

GROUP :—Chlorinated.

CLASS I :—*Sodio-chlorinated.*

Special applications :—Scrofula, lymphatism.

General applications:—Rheumatism, paralysis, surgical diseases, hæmorrhoids (abdominal plethora).

Secondary applications:—Dermatoses, hypochondria, dyspepsia, syphilis.

CLASS II:—*Chloro-sulphuretted.*

Applications, appropriate to both chlorinated and sulphuretted, but more specially to the latter:—Scrofula, dermatoses, rheumatism.

CLASS III :—*Chloro-bicarbonated.*

Applications :—Same as sodio-chlorinated.

CLASS IV :—*Chloro-sulphated.*

Application :—Speciality for dermatoses.

GROUP :—Bicarbonated.

CLASS I :—*Bicarbonated.*

Division 1 :—Sodio-bicarbonated.

Special applications :—Uric acid diathesis (gout, red gravel), obesity, diabetes, liver complaints, abdominal obstructions.

General applications :—Dermatoses, rheumatism, chronic metritis.

Divisions 2 and 3 :—Calcareous and mixed bicarbonated.

Application :—Dyspepsia (digestive table-waters).

CLASS II:—*Bicarbonato-chlorinated.*

Applications :—Same as sodio-bicarbonated and sodio-chlorinated, but with weaker effect.

CLASS III :—*Bicarbonato-sulphated.*

Applications :—Catarrhal affections of the urinary organs, dyspepsia.

CLASS IV :—*Bicarbonated mixed (sulphates and chlorides).*

Applications :—Same as sodio-bicarbonated, with the addition of laxative properties.

GROUP :—Sulphated.

Division 1 :—Sodio-sulphated.

Applications :—Complaints requiring laxative action.

Divisions 2 and 3 :—Calcareous and mixed (sodio-calcareous) sulphated.

Applications :—Analogous to the indeterminate (or simple thermal) waters.

Division 4 :—Magnesian sulphated.

Applications :—Laxative and purgative.

GROUP :—Indeterminate.

CLASS I :—*Simple thermal waters.*

Special applications :—General neuroses, neuralgia, rheumatism, dermatoses, chronic metritis.
General applications :—Neuropathic predominancy.

CLASS II :—*Feebly mineralised waters.*

Various applications :—Affections of the respiratory system, dermatoses, rheumatism, dyspepsia.

SUPPLEMENTARY GROUP :—Ferruginous Waters.

Applications :—Anæmia, chlorosis.

MODES OF ADMINISTRATION OF MINERAL WATERS

Mineral waters are administered in baths and internally, and also under various forms that may be called *accessory.*

Of these various methods the Bath is the essential one, and it alone was in use among the ancients.

Some mineral waters are used in no other manner, as, for example, those of Néris, Plombières, and Aix (Provence). These are waters of a feeble mineralisation, but of an effective and usually high thermality, flowing from sufficiently abundant springs to allow of their being largely used for bathing purposes.

The most perfect *thermal bath* is that which can be taken in running water, at a native medium temperature of between 28° and 36° C. (84° and 96° F.)   In dealing with high thermalities, or strongly mineralised waters, or certain pathological conditions, it is necessary carefully to limit the duration of the bath ; but in general it may be affirmed that the long duration of a mineral bath largely increases its efficiency ; and that, if it is necessary to reduce to a quarter of an hour the very hot baths of Mont-Dore, on the other hand, the patient may remain for several hours in the temperate baths of Luxeuil or Néris. For a prolonged bath, the *piscine* (or large public bath) which permits exercise in the water, is to be preferred to the other modes of bathing.

Mineral waters are also taken as drinks ; but their internal use occupies less place in thermal treatment than their external, although it is not of minor importance. The waters which are used externally are generally of a more notable mineralisation, and endowed with more distinct therapeutical properties ; but it is impossible to make any positive statement on the subject, for the manner of administering mineral waters necessarily depends upon the nature of the complaint treated, and of the indications which are to be obeyed. Although there are circumstances under which cold water is alone tolerated, yet on the whole it may be said that a temperature approaching that of the blood affords the best applications.

It is important that the amount of the *dose* should be fixed with great circumspection, and it should be borne in mind that all patients, and sometimes even doctors, have a tendency to raise doses beyond what is really required by necessity, or recommended by prudence. The drinking-fountains used by patients at the mineral springs are called in France *buvettes*.

Douches and *inhalations* are the accessory methods of administration which, under certain circumstances, play a considerable and very important part in the treatment.

The use of *douches* depends much more on the indication to be fulfilled than on the quality of the water used, and the percussion and the temperature have a much more important action than the constituents of the mineral water. Douches are of two classes, local or *resolutive,* and general (as they may be termed) or *revulsive,*—the former intended to act upon some unhealthy organ, and the latter to affect the system in general.

Besides the ordinary downward stream of water, *ascending douches* are also used. These are the *rectal* or simply *anal*. They can be employed as evacuating, resolutive, or revulsive.

Vaginal douches should only be used with the utmost

caution, and under the special advice of a physician, so that their use should be strictly limited to their legitimate applications. Simple vaginal *irrigations* are much more generally and safely employed. In these last, the quality of the mineral water has a more therapeutical value than in the external or percussion douches, and the temperature must receive special attention.

Inhalation is a very interesting accessory method of administering mineral water. Its object is to introduce into the pulmonary organs either special vapours or gases. Inhalations are chiefly used in connection with sulphureous waters, but are also practised at mineral springs of other classes, such as those of Mont-Dore, La Bourboule, Royat, &c. In the case of sulphurous waters, which give off sulphuretted hydrogen on contact with the air, spontaneous inhalation is inseparable from all modes of administration, both external and internal. But inhalation is also practised there systematically, either by breathing the thermal waters themselves, mixed as they naturally are with sulphuretted hydrogen, or else breathing the gas alone, its liberation having been promoted by the breaking up of the mineral water.

Pulverisation, which was introduced into hydrothermal practice by Dr Sales-Girons, consists in converting the mineral water itself into the form of very dense powder or smoke, so that it can be inhaled in its integrity. But as the pulverised water penetrates only very imperfectly into the bronchial tubes, it must be used in affections of the upper portions of the respiratory apparatus, while gas and vapours must be reserved for the deeper applications.

The best methods of inhalation are those which enable the patient to breathe the gas or vapours, without being plunged into a humid atmosphere, which would transform the inhalation-room into a hot vapour-bath.

Mineralised mud-baths are used under certain circumstances, especially in chronic rheumatism, chronic rheumatic arthritis, old traumatism, or partial muscular atrophy. They are formed of earth saturated with certain mineral waters, and impregnated with gaseous or saline principles. The salts generally found in these *boues* or mud—are those of lime, magnesia, iron, and sulphuretted hydrogen arising from the decomposition of sulphates. The formation of this therapeutical mud seems rather to be favoured by the nature of the soil than dependent on the composition of the waters which penetrate it.

Special mention must also be made of *carbonic acid gas*, which can be used at the springs impregnated with it, either in inhalations, baths, or injections. This mode of treatment was first employed at Saint-Alban (Loire), and is now very completely established at Vichy, as also at Saint-Nectaire and Royat. The baths form the most interesting part of the treatment. When thus used, carbonic acid acts at the same time as a sedative to pain, and as a stimulant to the innervation, particularly to the development of natural heat.

It seems to act especially upon the subcutaneous net-work of nerves and capillaries. Carbonic acid baths are most clearly indicated in cases of painful neuroses, perversions of movement, atony, or loss of natural heat.

GEOGRAPHICAL DIVISION OF THERMAL STATIONS

France is certainly one of the most fortunately endowed of countries in the matter of mineral waters, and even more so in respect of their variety—which offers the most remarkable types of all mineralisations, combined with the most widely differing temperatures —than of their absolute number.

Out of about 1000 mineral springs which have been noted in France, at least 800 belong to the moun-tainous regions, and proceed from rocks of an igneous origin, or sedimentary deposits which bear more or less profoundly the traces of their action.

If we go farther, and examine attentively the pre-dominating characteristics of the waters of such or such a mountainous district, we shall soon perceive that there again preferences exist; and it would not be difficult to show, for instance, that acidulous waters are as abundant in the central region of France as sulphureous springs in the chain of the Pyrenees. On the other hand, the plain country, sheltered as it is from abnormal influences, usually offers only springs which—from their low temperatures, and the nature of the elements that they hold in solution—can generally be considered as the result of a process of filtering (carrying with them, probably, a certain quantity of carbonic acid) through the superficial strata.

The geographical or *regional* distribution of the mineral springs of France possesses, therefore, a double interest, from the very fact of their unequal partition, and also of the preponderance of different elements in the waters of different regions.

It is with the object of facilitating the study of the different groups that we have decided on dividing the Thermal Stations of France into six great divisions, as follows :—

> I. Central Division,
> II. Pyrenean Division,
> III. Southern Division and Corsica,
> IV. Eastern Division and Savoy,
> V. Northern Division,
> VI. Western Division.

It is necessary to premise that in the pages which follow, we have (with few exceptions) noticed only those Thermal Stations which, by the therapeutical value of their springs, and the adequacy of their arrangements for the reception of visitors, can offer to persons in search of health a comfortable and agreeable residence.

If we have given, so to speak, the place of honour in this book to the springs of the centre of France, it is because—for several reasons—they offer greater advantages to patients than those of the other divisions. The Stations are more easily accessible, the therapeutical application of the waters is more general and more extended ; and, finally, their geographical position, the completeness of their arrangements, and the reputation which they have acquired, confer on them a pre-eminence over the springs of the other regions which it has been necessary to recognise.

THE

MINERAL WATERS OF FRANCE

CENTRAL DIVISION

THE mineral waters of the centre of France are chiefly to be found among the mountains of Auvergne, or in the ancient province of the Bourbonnais. The number of the springs is very large, and their temperature usually high; in some cases, indeed, it approaches the boiling point. While in the Pyrenees sûlphureous waters abound, the centre of France scarcely possesses any; but in exchange it can boast of saline waters of the highest rank. The salts which mineralise them are especially bicarbonates, sulphates, and chlorates; the dominating base is soda; and, finally, the gas which most abounds—sometimes even to saturation— is carbonic acid.

ROYAT

(*Puy-de-Dôme*)

MIXED SODIO-BICARBONATED SPRINGS, CONTAINING IRON
AND ARSENIC.

ROYAT is a village with a population of about 500,
a little over a mile from Clermont. It is built at a
height of nearly 1500 feet above the level of the sea,
between two mountains covered with rich vegetation,
at the entrance of a deep ravine scooped out by a
torrent of lava. We cannot improve on the following description, given by M. Eugène Guinot: "Its
white houses, mills, and cottages, dotted about a gentle
slope, gleam out among the trees as from a nest of
verdure. Above rises the church, of an imposing aspect,
adorned with towers and battlements, like a fortress.
Below the village is the celebrated grotto of Royat with
its springs, which gush out in cascades, and flow away
to join the Tiretaine stream. Lovers of the beauties of
Germany and Switzerland will find nothing in their
sketch-books more picturesque or more softly beautiful
than the landscape formed by these rocks, these woods,
these cascades, this village clinging to the mountain
and smiling through its shady trees, this stern-looking
church, and this marvellous grotto which seems the
cool and mysterious retreat of some mythological
divinity."

The Thermal Establishment, one of the finest and
most complete in France, stands in the most picturesque

part of the valley, between the right bank of the Tiretaine and the road which leads from Royat to Mont-Dore. It includes no fewer than ninety-four bath-rooms, with baths of Volric stone, each provided with apparatus for local douches, vaginal, articular, &c. It has, besides, independent apparata for douches of various forms, two inhaling-rooms, a large public swimming bath traversed by a current of fresh water, which is reserved for ladies in the morning, and for men in the afternoon; arrangements for carbonic acid baths and douches, as well as for the water cure system, and lastly, a gymnasium.

The waters of Royat have been in use since the year 1845, and its Thermal Establishment was declared of public utility in 1860. The springs are four in number, and the following table gives their names, temperatures, and daily product.

Name of spring.	Temperature.		Product in twenty-four hours.
Grande Source, or Source Eugénie	35·5° C.	96° F.	317·520 gallons.
Source César	29° C.	84° F.	79·380 ,,
Source Saint-Mart . . .	31° C.	88° F.	162·288 ,,
Source Saint-Victor . . .	20° C.	68° F.	46·305 ,,

The springs which supply the establishments of the baths of Cæsar and of Saint-Mart rise on the left bank of the Tiretaine stream, and the Saint-Victor spring upon the right. The water of all is clear and colourless, almost without smell, of a sharp flavour something like that of ink, with an alkaline after-taste.

CHEMICAL ANALYSIS

WATER 1000 grammes (1 litre).	GRANDE SOURCE, OU SOURCE EUGÉNIE.	SOURCE CÉSAR.	SOURCE SAINT-MARTIN.	SOURCE SAINT-VICTOR.
Carbonic acid, free Litre	0·377	0·620	0·532	0·831
	Grammes.	Grammes.	Grammes.	Grammes.
Bicarbonate of soda . .	1·349	0·392	0·421	0·428
„ potash .	0·435	0·286	0·365	0·312
„ lime .	1·000	0·686	0·953	0·822
„ magnesia	0·677	0·397	0·611	0·514
„ iron .	0·040	0·025	0·042	0·042
Sulphate of soda . . .	0·185	0·115	0·163	0·123
Phosphate of soda . .	0·018	0·014	0·007	0·005
Chloride of sodium .	1·728	0·766	1·682	1·165
Silica	0·156	0·167	0·102	0·089
	5·724	4·067	5·396	5·146

Traces of bicarbonate of manganese, arseniate of soda, iodide and bromide of sodium, alumina and organic matter.—*Lefort.*

From this analysis, the latest that we possess, it appears that the bicarbonates of soda, lime, and magnesia, together with chloride of sodium, and bicarbonate of iron with traces of arsenic, are the predominating components of the Royat water. It is therefore distinctly alkaline, and may be classed among the mixed bicarbonated and chlorinated waters.

This medium mineralisation of the Royat waters points to a medication which is not confined in its applications only to well-defined constitutional conditions; but there is also a certain variety of *lymphatism,* anæmia, and nervous disorganisation, which it perfectly suits, and which is to be met in connection with rheumatism, chlorosis, uterine complaints, dyspepsia and gastralgia; in all cases the complex character of the malady being the indication that the Royat waters are likely to be beneficial.

These waters are stimulating, diuretic, laxative, tonic, and restorative. They are used as drinks, for baths and douches, and for inhaling. Administered in the latter way, they are a powerful therapeutical agent, affecting at the same time the external integument and the pharyngeal and pulmonary mucous membranes.

Bearing in mind the qualifications mentioned above, the complaints for which the Royat waters may be recommended fall into three great categories.

1st. *Affections of the respiratory passages.*—Under this head are included laryngitis and chronic bronchitis, pulmonary catarrh, emphysema and humid asthma, for which the water is taken both externally and internally, but chiefly by inhalation. This latter mode of administration is certainly the best, and ought to supersede the others, since everyone now knows how powerful a sedative effect is produced upon the respiratory organs by the absorption of warm mineral vapours in conjunction with a large quantity of carbonic acid.

2nd. *Nervous affections.*—In this category are comprised: chloro-anæmia, so frequent among persons whose nerves are out of order; neuroses, and the numerous forms of dyspepsia, (whether of the stomach or the intestines), as well as their originating causes, which in the case of women are usually uterine diseases and congestions, leucorrhœa, amenorrhœa, and dysmenorrhœa,—and in the case of men are over-study, want of sleep, anxiety; in short, all the causes which may produce poverty of blood.

3rd. *Cutaneous affections.*—Under this third category may be included the different manifestations produced by the influence of *arthritis*, which M. Bazin defines as "a constitutional malady, characterised by various manifestations in the different systems of the organism, and especially by affections of the skin, symptoms in the joints, and the tendency to the formation of a morbid product, the tophus." These manifestations are gout, rheumatism, impetigo, acne, and eczema (which alone contributes four-fifths of the cutaneous cases treated at Royat).

The discovery that lithine (chloride of lithium) existed in considerable quantities in the Royat water has for the last few years attracted to this Station a large number of gouty patients. The remarkable experiments made in England by Dr Garrod, and repeated in France by Dr Charcot, and many other practitioners, have demonstrated the property which lithia possesses in a higher degree than any other alkaline substance, of combating the morbid action that generates an excess of uric acid in the system of persons suffering from gout, and of destroying the tophaceous products which it deposits around the joints.

But the patients who derive most benefit from the treatment at Royat are those in whom the gouty diathesis, while affecting the digestive functions, has already produced in the system a cachectic atony which often resists all that can be done by the most judicious treatment.

M. Rotureau compares these waters to those of Ems from several points of view, and considers that Royat can with advantage be substituted for the German watering-place, for whatever reason prescribed.

Persons suffering from acute affections ought not to use the Royat waters, nor those affected with cancer, aneurism, softening of the brain or spinal cord, or acute phthisis.

Connected with the Thermal Establishment is one for hydropathic treatment. It is divided into two corresponding sections, one for men, and the other for women. Each contains a large public bath and douche, a hot-air room, a small room for foot-baths and hip-baths, and a dressing room. Besides the cold water treatment, " Scotch," Tivoli, and mineral baths, both cold and graduated, can be administered in the same establishment, and it contains also a Russian bath.

Pretty gardens surround the whole, which is further completed by the addition of a well-arranged gymnasium.

The season lasts from May 15th to September 15th.

The waters keep perfectly, and are exported without undergoing any deterioration.

The description quoted at the beginning of this chapter is enough to show that Royat is in itself a charming place to visit; but there is also much besides the beauties of nature to make a stay there interesting, as many curiosities, both natural and artificial, are to be found in the neighbourhood. Among the most remarkable may be mentioned the beautiful Gothic cross, erected in the public square, on which are sculptured the twelve Apostles; the Church, of a Romano-Byzantine style of architecture, dating from the twelfth century; the famous grotto of basaltic rocks, from which spring several inexhaustible fountains; the grotto of the thermal springs; the village of *Fontenel*, &c.

The most interesting excursion is that to the *Puy-de-Dôme*, where there is an observatory, and also some ancient ruins; and whence, on a clear day, an immense and magnificent panorama can be enjoyed. Clermont can be seen immediately beneath, while beyond lies all the chain of the Monts-Dômes, the Monts-Dores, and the mountains of Cantal.

SAINT MAURICE, or Vic-le-Comte. (*Puy-de-Dôme.*) —The temperature of this water is from 16° to 34° C. (61° to 93° F.). It contains per litre 4 grammes of mixed bicarbonates, 2 grammes of chloride of sodium, and 0·04 gramme of bicarbonate of iron. This is a very interesting composition, approximating to the group of applications given for the Royat water, and admitting of rivalry with the chlorinated carbonic waters of Germany, so much praised for their effects in disorders of the digestive apparatus.

MONT-DORE

(*Puy-de-Dôme*)

SODIO-BICARBONATED, FERRUGINOUS, AND ARSENICAL
SPRINGS, FEEBLY MINERALISED.

THE valley of Mont-Dore is one of the most curious
and picturesque parts of old Auvergne. The elevation of the strata, the craters, and beds of lava proves
that in distant ages this country, now so peaceful
and fertile, was upturned by frightful cataclysms.
The village of the baths is situated in the valley,
through which flows the Dordogne, which at this
point is only a stream, almost dry in summer, and, as
it were, lost in a rocky ravine. It is on the right
bank, at the base of the Angle mountain, that the
mineral waters have their source.

This watering-place was known to the Romans, as
will be readily recognised by any one who walks
along the promenade belonging to the Thermal Establishment. This is adorned with remains of columns,
capitals, sepulchral stones, and different fragments of
statues in porphyritic lava, the most interesting of
which is a well-preserved and remarkably expressive
head of Nero. All these objects have been taken
from the ruins of a Temple and Thermal baths which
were discovered in 1825, in the course of excavations
carried on in the old square known as the *Panthéon.*

The waters proceed from eight springs, of which seven are hot and one cold. They supply two very well-conducted Establishments, one reserved for baths and douches, the other for inhalations and vapour douches.

The thermal springs, known under the names of the *Madeleine* (or *Bertrand*), *Pavillon* (or *Grand Bain*), *Ramond, Rigny, Boyers,* and *Pigeon,* have a temperature varying from 42° to 45° C. (108° to 113° F.) They give daily more than 77,161 gallons of a water clear and colourless at the source, but which grows turbid on contact with the air, and becomes covered with oily drops, which gradually spread and soon form a thin iridescent layer. The water has no smell; its taste is at first slightly acid, but afterwards salt, with a lixivial after-taste.

The cold spring (temperature 12° C., 54° F.), known under the name of the *Fontaine Sainte-Marguerite,* gives daily more than 5500 gallons of a somewhat turbid water, of an acid and piquant flavour, leaving a bitter after-taste, and giving off much carbonic acid gas.

As will appear from the analysis, the Mont-Dore water does not belong to the class of the highly mineralised. And here it may be well to observe that the physiological properties of any water cannot be deduced from the quantity of mineral salts which it contains, and that it is often impossible to establish any proportion between the degree of mineralisation and the therapeutical effects.

CHEMICAL ANALYSIS

WATER 1000 grammes (1 litre).	SOURCE CÉSAR.
	Grammes.
Oxygen	0·98
Nitrogen	14·22
Carbonic acid, free	0·5967
Bicarbonate of soda	0·5361
„ potash	0·0212
„ lime	0·3209
„ magnesia	0·1676
„ and oxide of iron . . .	0·0558
Chloride of sodium	0·3587
Sulphate of soda	0·0756
Arseniate of soda	0·00096
Silicic acid	0.1552
Alumina	0·0083
Traces of bicarbonates of rubidium, of lithia, of oxide of caesium, and of manganese, borate of soda, iodide and fluoride of sodium, and bituminous organic matter.—*Lefort.*	2·29706

The treatment at Mont-Dore is based principally upon the "thermality" of its waters, and consists of short baths at a temperature of from 40° to 45° C. (104° to 113° F.), and drinks equally hot. Here we meet—in connexion with a very feeble and slightly characterised mineralisation—phenomena which seem to belong to the active sulphuretted or strong chlorinated waters, namely, stimulation of the skin and profuse perspiration, return of neuralgic, rheumatic, or gouty pains, or of cutaneous affections. The greater part of the effects of the water seem due principally to its "thermality," which could scarcely be utilised if it were more strongly mineralised.

But what completely distinguishes these waters

from the sulphurous class is that this peripheral stimulation ends in sedative and hyposthenic effects, so that the Mont-Dore is particularly suited to irritable cases which cannot bear sulphurous waters, and unsuited to the torpid cases for which the latter are beneficial.

The waters are administered in three principal ways at Mont-Dore, as drinks, inhalations, and baths. But the conjunction of all three methods is seldom either necessary or possible. Inhalation occupies an important position in the treatment. The two Establishments include ninety-six bath-rooms, two public baths, two rooms for foot-baths, seventy-two descending and two ascending douches, two naso-pharyngeal douches, eight inhaling rooms, two pulverisation rooms, twenty-two vapour douches, and eight vapour baths.

The waters are generally drunk very hot, the temperature of the *Bertrand* and *Ramond* being from about 42° to 45° C. (108° to 113° F.) The dose is two or three glasses a day. They are quickly absorbed by the stomach, stimulate the appetite, and greatly quicken the circulation.

The diseases, which are most largely to be met with at Mont-Dore, are chronic affections of the respiratory organs, against which their assimilating action is truly remarkable. Chronic bronchitis, asthma, chronic bronchorrhœa, chronic laryngitis, and pharyngitis and aphony are treated with great success.

Consumptive patients also resort to Mont-Dore,

and the action of the waters on such cases is thus explained by M. Richelot:—"In the consumptive patient, under the influence of the Mont-Dore waters, nervous energy increases, the consuming rapidity of the circulation abates, the fever disappears, the inflammatory congestion of the portions of pulmonary tissue surrounding the tuberculous deposit tends to disperse, respiration becomes deeper, the circulation in the lungs less incomplete, hæmatosis is more easily accomplished; the appetite increases, nourishment is more easily digested, and without fatigue to the organs or feverish reaction; general nutrition makes rapid progress, strength revives, and the patient begins to gain flesh. If the treatment is employed when consumption is in its earliest stage, it may act up to a certain point as a prophylactic agent, and prevent the development of the malady by restoring the vital forces. If, however, the disease be far advanced, the treatment can, at least, check its progress, and prolong the life of the patient."

The general arrangements for the use of the Mont-Dore waters are far from corresponding to the therapeutical importance of the remedy, and it is to be regretted that they still remain so very primitive. It may be gratifying to an antiquary to bathe in a bath which has been used by the ancient Romans; but the fragments of enormous columns and cornices belonging to the old *thermæ*, which have been discovered on the site of the present baths, would lead to the supposition that only their third class Establishment has survived.

A little modern comfort would not detract from the efficacy of the waters, patients would profit by the improvement, and the locality would be the gainer.

The climate of Mont-Dore is a mountain climate; the mornings and evenings are cool, or even cold, and invalids are often obliged to put on their winter clothing after sunset.

The thermal season lasts from the 1st of June to the end of September, but the best time to visit Mont-Dore is between July 1st and August 30th.

These waters can be exported, and kept for a long time without losing their qualities.

The walks and excursions in the neighbourhood are charming. Among the most attractive will be found those to the *Pic-de-Sancy*, the *Puy-Ferrand*, the sources of the rivers *Dore* and *Dogne*, the valley of *Chaudefour*, the *Capucin*, the *Salon de Mirabeau*, the lakes of *Pavin*, *Guéry*, and *Chambon*; the cascades of *Mont-Dore*, *Plat-à-barbe*, and *Le Serpent*, the rocks of *Tuillère* and *Sanadoire*, &c.

LA BOURBOULE

(*Puy-de-Dôme*)

SODIO-CHLORINATED, BICARBONATED, AND ARSENICAL SPRINGS.

LA BOURBOULE is a small village 2790 feet above the sea level, at the foot of an immense granite rock, in a smiling valley watered by the little river Dordogne.

The mineral waters of La Bourboule are derived from two springs known as the *Source Perrière* and the *Source Choussy.*

Both of these are drawn from artesian wells. They issue from the granite at a depth of from 245 to 260 feet. Two large pumps raise the water to the surface. On issuing from them it has a temperature of 54° C. (130° F.), but at its source it is 60° C. (140° F.)

The water is employed as a drink, and for baths, douches, pulverisation, inhalations of vapour, &c.

There are also several other springs equally drawn from artesian borings, but they are of secondary importance. We have, therefore, only to consider the two already mentioned.

The wells from which they are drawn are close together, as are also their subterranean channels. Apparently they communicate with each other, and it is probably the same water which is drawn from each pump, though under different names.

The "Bourboule Mineral Water Company" manages the three Thermal Establishments, *Les Thermes, Choussy,* and *Mabru.*

The *Etablissement des Thermes,* which is reserved for the use of patients of the upper classes, is situated in the centre of the village, on the banks of the stream. It consists of a large rectangle, at each corner of which is a pavilion covered by a dome. Each pair of pavilions is connected by a handsome gallery, which passes through another pavilion in the centre of the grand façade. Opening from these galleries are the bath-

rooms, those in the one being for men, and in the other
for ladies. In the corners of the end pavilions are
the inhalation-rooms, the vapour-rooms, ascending
douches, and dressing-rooms. The smaller sides of the
rectangle contain the pulverisation-rooms, a vapour-
room for men and one for ladies, foot-bath rooms, and
dressing-rooms. The central pavilions, which contain
the offices, &c., are connected by a very large gallery
used as a conversation-hall; and in the middle of this
is the *buvette*. On each side of this gallery are the
" grand douche " rooms and those for hydro-therapeutics and shampooing. Each bath-room contains an
enamelled cast-iron bath with douche apparatus.
The water enters at a considerable pressure, coming
from a height of thirty-five feet. Several specially
luxurious bath-rooms have private salons attached :
the others open out of a room fitted up as a dressing-
room. The pulverisation-rooms are very large, and
each contains twenty-one douches fitted with the most
complete appliances.

In the " grand douche " rooms are very complete
and perfect appliances for horizontal douches, circular
douche, shower douche ; sitting, dorsal, and vaginal
douches.

The *Choussy* Establishment is more simple, but
is large, and furnished with everything which the
present state of hydro-therapeutics renders necessary.

The *Mabru* Establishment, which has been recently
reconstructed, although plain, is equally well cal-

culated to give satisfaction to its visitors, who are principally patients sent by charitable institutions.

CHEMICAL ANALYSIS

WATER 1000 grammes (1 litre).	SOURCE PER-RIÈRE, OU DU GRAND-BAIN.
	Grammes.
Carbonic acid, free	0·3852
Chloride of sodium	3·3457
„ potassium	0·2353
„ magnesium	0·0390
Sulphate of soda	0·2788
Bicarbonate of soda	2·2719
„ lime	0·1964
Arseniate of soda	0·01263
Silicic acid	0·1093
Alumina	0·0301
Traces of chloride of lithium, of cæsium, and of rubidium, bicarbonate of iron, of ammonia and of manganese, phosphate of soda, iodide and bromide of sodium, and bituminous organic matter.—*Lefort.*	6·90433

Clear, colourless, and inodorous, of a taste first acid and then alkaline, La Bourboule waters give off a considerable quantity of carbonic acid gas. It results from various analyses that they contain arsenic in larger quantities than any other mineral water (twenty-eight milligrammes of arseniate of soda to the litre), accompanied by two adjuncts, bicarbonate of soda, and chloride of sodium, which play a double part in conjunction with the arsenic, causing it to be tolerated by the most delicate stomachs, and aiding powerfully to ensure its beneficial effects. One litre (1000 grammes) of La Bourboule water is equal to twenty-one drops of Fowler's Solution.

The complaints for which these waters are pre-scribed are :—Anæmia, lymphatism, and scrofula (even its most deep-seated forms, caries and necrosis of the bones, being treated there with advantage, when once the acute period is passed) : chronic rheumatism, with enlargement of the joints; diabetes and albuminuria may also be added to the list. It has been established by decisive observations that diabetic patients, on whom the waters of Vichy and Vals had produced no effect, have improved remarkably under the treatment at La Bourboule. Certain skin diseases (such as her-petic affections, eczema, psoriasis, &c.) are also treated there with success; and the same may be said of affec-tions of the respiratory organs : chronic laryngitis and bronchitis, granular pharyngitis, asthma, emphy-sema, &c.

The name of *Source des Fièvres*, formerly given to one of the springs of La Bourboule, bears witness to the beneficial properties which these mineral waters have always possessed for the cure of intermittent fevers.

La Bourboule waters are unsuitable in all cases in which hæmoptysis is to be apprehended, in affections of the heart and large blood-vessels, and when there is a distinct apoplectic tendency.

Persons intending to visit La Bourboule ought not to forget, when making their preparations for the journey, that they are going to a mountain country, elevated 2788 feet above the level of the sea, and exposed in a certain degree to the variations of temperature which always belong to such an altitude. Thus, although the climate of La Bourboule is (comparatively speaking) very mild, thanks to the high wall of granite which protects it from the north wind, it would not be wise to go there unprovided with warm clothing; and sensitive persons, especially the rheumatic, ought always to wear woollen material next to the skin.

The thermal season begins on June 1st and ends on September 15th.

La Bourboule is a centre whence can be made many most interesting and varied excursions, of which the greater part can be enjoyed without any fatigue, on foot, or on horseback, or on donkeys. Among the most attractive may be mentioned :—The rock of *Vendeix* and the forest of *Bozat*, the villages of *Saint Sauves* and *Liornat*, the cascades of *Plat-à-Barbe* and of *La Vernière*, the *Salon de Mirabeau*, the cascades of *Queureilh*, *Rossignolet*, and *Le Serpent*, the valley of *La Cour*, and the *Gorge d'Enfer*.

SAINT-NECTAIRE

(*Puy-de-Dôme*)

SODIO-CHLORINATED, BICARBONATED, FERRUGINOUS, AND GASEOUS SPRINGS.

SAINT-NECTAIRE is a town of 1500 inhabitants, built on a hill, at a height of 2500 feet above the level of

4

the sea.   It stands among charming scenery and wonders of nature, whose beauty must satisfy the most
exacting tourist.   It is divided into two parts, Upper
and Lower Saint-Nectaire.

There are four Establishments, called respectively
the *Bains Romqins* (or *Mandon*), *Bains Boëtte*, *Bains
du Mont-Cornador*, and *Bain Chanzède*.   The first
three are much the most important, and their hydro-
therapeutical arrangements are very complete.   They
are supplied by a considerable number of springs of
almost identical composition, and of temperatures
varying between 18° and 46° C. (64° and 115° F.)
Their aggregate product is 44,000 gallons in the
twenty-four hours.

The most important springs are six in number. They
are named as follows :—*Source Boëtte, Source Cornador.
Source Mandon tempérée, Source Mandon chaude, Source
Pauline,* and *Source Rouge.*   They yield a clear water,
soft and oily to the touch, of an hepatic smell, and a
taste at first acid, but afterwards alkaline and ferruginous.   It soon loses its transparency on contact with
the air, takes a dim hue, becomes covered with a crystal-
line coating of a yellowish white colour, and deposits
at the bottom of the reservoirs an ochreous sediment.

It will appear from the analysis that the waters of
Saint-Nectaire are some of the most remarkable in
respect of their mineralisation, containing, as they do,
more than seven grammes to the litre of alkaline
principles ; bicarbonate of soda, and chloride of sodium
being present in almost equal quantities, as well as

iron, and free carbonic acid gas. They are, therefore, bicarbonated and sodio-chlorinated, ferruginous, and gaseous waters.

CHEMICAL ANALYSIS

WATER 1000 grammes (1 litre).	SOURCE MANDON.
	Grammes.
Carbonic acid, free	1·5308
Chloride of sodium	2·4148
Bicarbonate of soda	2·0881
„ potash	0·0407
„ lime	0·7060
„ magnesia	0·4815
„ and oxide of iron . . .	0·0097
Sulphate of soda	0·1718
„ strontia	0·0070
Alumina	0·0205
Silica	0·1036
Traces of oxygen, nitrogen, iodide of sodium, arseniate of soda, phosphate of soda and bituminous organic matter.—*Lefort.*	7·5808

Taken internally, these waters are at the beginning of the course of treatment stimulating to the digestive functions. In doses of two or three glasses per day, they increase the appetite, excite thirst, and induce constipation; they are also diuretic, and render the urine alkaline. In larger doses, they become laxative, or even purgative.

Baths constitute the most important part of the treatment at Saint-Nectaire. They produce a marked excitement of the nervous system in general, while by means of the carbonic acid which they contain, they act as an anæsthetic upon the skin. On the whole, they are stimulating, tonic, and restorative, in consequence of their carbonic acid, chloride of sodium, iron, and arsenic; and at the same time resolutive and antiplastic, in consequence of their bicarbonate of soda.

In the first rank of the affections to be met with at Saint-Nectaire, must be placed, chronic, muscular, and articular rheumatism. Next to this come the group of neuralgias, and especially sciatica; chronic catarrhal affections of the uterus; inflammatory cutaneous manifestations of scrofula and arthritis; chlorosis, and swellings accompanied with stiffness of the joints, consequent on fractures. Lastly, M. Dumas-Aubergier has lately made successful use of these waters in the form of pulverisation, for the treatment of granulous conjunctivitis, chronic blepharitis, and opacity of the cornea.

The season commences on the 1st of June, and terminates in the beginning of September.

The first impression which Saint-Nectaire produces upon the newly-arrived visitor is not very agreeable; but when he has spent some days in exploring the place and the neighbourhood, he will admit that this Station is surrounded by very picturesque bits of landscape, and after his departure he will retain a pleasant remembrance of his excursions, which will have proved both enjoyable and interesting.

In speaking of Saint-Nectaire, it is impossible to leave unnoticed the grotto where the petrifying springs rise to the surface. The inhabitants carry on an extensive trade in petrified objects; and the visitor will

have his choice among a collection of the most varied
articles, from the coarsest and simplest to the finest
cameos. The latter have been produced by the ingenious device of using gutta-percha moulds.

VIC-SUR-CÈRE

(Cantal)

GASEOUS, SODIO-BICARBONATED, AND FERRUGINOUS SPRINGS.

At the foot of the Cantal range, and at a distance
of twelve miles from Aurillac, rise the mineral
springs of Vic-sur-Cère. This is a highly gaseous
water, mineralised by alkaline and ferruginous salts :
3·109 grammes of the former, and 0·050 grammes of
the latter, to the litre. Its composition is exactly
the same as that of the natural Seltzer waters, but
richer in iron and in bicarbonate of soda. It will be
found suitable wherever tonic or restorative treatment
is required for cases of general debility.

CHATEL-GUYON

(Puy-de-Dôme)

SODIO-CHLORINATED, BICARBONATED, FERRUGINOUS AND
GASEOUS WARM SPRINGS.

Chatel-Guyon is a small town, containing nearly
2000 inhabitants, about four miles from *Riom*, in one

of the most delightful parts of Auvergne. It stands
at an elevation of 1300 feet above the level of the sea.
Its mineral waters proceed from several springs rising
on both banks of a stream called *Sardour*. Their
temperature varies between 28° and 35° C. (82° and
95° F.) Their aggregate daily product is about
154,000 gallons, but they are not all utilised.

There are two Thermal Establishments; the oldest
—the *Établissement Barse*—only contains a few private
baths, and two *piscines* most unsuitably placed. The
other, the *Établissement Brosson*, of recent construction, unites the greater part of the advantages which
the visitor to a Station of this importance has a right
to expect. It contains twenty-two baths, two large
piscines in which the water is constantly renewed,
twenty bath-rooms, in which are apparata for every
possible variety of douche,—and lastly, two special
rooms for ascending and one for vaginal douches.

The mineral springs are fourteen in number, of
which the following are the chief; *Source Deval,
Source Pardon, Source Barse, Source du Sopinet,
Source de la Planche, Source du Réservoir, Source de la
Vernière*, &c.

The water of all these different springs is transparent
when it issues from the ground, but after having been
exposed to the air assumes a slightly opaline hue. It
has no smell; its taste is somewhat sharp, with a
rather salt after-taste. According to the most recent
analysis, it contains six grammes of salts to the
litre.

CHEMICAL ANALYSIS

WATER 1000 grammes (1 litre).	SOURCE DEVAL.
	Grammes.
Carbonic acid, free and combined . . .	2·142
Hydrochloric acid	2·133
Sulphuric acid	0·293
Silicic acid	0·126
Potash	0·112
Soda	1·287
Lime	0·990
Magnesia	0·670
Alumina	0·008
Oxide of iron	0·024
Traces of arsenic acid, strontia, lithia and bituminous organic matter.	8·085

These are, as it is obvious, mixed waters, the composition of which partakes of the chlorinated, alkaline, bicarbonated, ferruginous, and lithinated characters. They also contain considerable quantities of free carbonic acid gas.

When the patient comes out of the bath, one can judge, by the quantity of fine bubbles which cover his whole body, of the considerable proportion of gas contained by this mineral water. The formation of these little bubbles on the surface of the skin, with the reaction that follows it, explains how the bath at first gives a refreshing feeling, soon followed by an opposite sensation, and thus is at once soothing and tonic.

The waters of Chatel-Guyon are digestive and laxative when taken in small doses, and purgative in larger ones ; they are also diuretic. These qualities make them especially valuable.

The affections for which they are most successfully prescribed are dyspepsia, flatulence, and gastrorrhœa (a true catarrh of the digestive organs), especially when accompanied by constipation. In such cases a few glasses of the mineral water taken in the morning, fasting, are enough to restore the digestion to its natural condition. Enlargements of the liver and spleen, and mesenteric obstructions, profit equally by the same treatment. The waters are also beneficial in enlargement and other chronic diseases of the uterus; and in cases of biliary calculi, jaundice, gravel, albuminuria, and cachectic diabetes. Lastly, by attracting the blood to the intestines, they remove the tendency to cerebral congestion which so often becomes the cause of paralytic attacks.

The comparison of the water of Chatel-Guyon to that of Pullna as a purgative is unfair to the former. It should rather be compared, if not preferred, to those of Carlsbad and Marienbad. It is of small consequence that it derives its valuable properties from chlorides and not from sulphates of soda, like the German waters, if it gives results equally perfect under the same pathological conditions. Its "thermality" constitutes a great advantage over Marienbad and Kissingen. Finally, Chatel-Guyon is only half a day's journey from Paris, and offers such advantages of climate and situation that it cannot fail one day to surpass its rivals as a watering-place.

The waters keep well, and are exported in considerable quantities.

ROUZAT

(*Puy-de-Dôme*)

MIXED BICARBONATED AND FERRUGINOUS SPRINGS

ROUZAT is a small Thermal Station situated on the territory of the parish of Beauregard-Vaudoy, at rather more than four miles from Riom. The mineral springs are two in number, and yield together a daily product of about 66,000 gallons of water, at a temperature of 31° C. (88° F.)

The iron, bicarbonate of lime, and chloride of sodium, which are found in considerable quantities in the waters of Rouzat, render them very useful in the treatment of chronic rheumatism, scrofula, and anæmia.

———

CHÂTEAUNEUF

(*Puy-de-Dôme*)

MIXED, BICARBONATED, FERRUGINOUS, AND GASEOUS SPRINGS.

CHÂTEAUNEUF is a little town with a population of about 1200, situated about fifteen miles from Riom, at a height of 1250 feet above the level of the sea. It is built on both banks of the river *Sioule*, in the midst of the picturesque mountains of Auvergne. At the entrance to this valley, the scenery is wild and dreary, and the eye rests only upon arid mountains, bristling with *aiguilles* of granite and porphyry. But as the traveller advances, the prospect changes, and the Valley of Châteauneuf soon appears in all its beauty. The bareness of the mountain slopes gives way to the dark green shade of the woods which clothe the heights, and descend to meet the green meadows of this delightful valley.

The mineral waters flow from no fewer than fifteen springs, which all seem to have a common origin. They are sodio-bicarbonated and ferruginous. Their temperature varies from 15° to 38° C. (59° to 100° F.), and their mineralisation from 3·487 gram. per litre to 6·756 gram.

These springs are too numerous to be studied individually, but they may be divided into three groups.

1st. *Ferruginous and gaseous cold springs.*—If we use the name ferruginous to distinguish this group of springs, it is not intended to imply that the others are destitute of iron; but that in this first group it is present in such abundance as to impress upon it a peculiar character.

Under this category come the *Source Morny, Source du Petit-Rocher,* and *Source du Petit-Moulin.* They show on analysis a proportion of from 50·42 to 62 milligrammes of carbonate of iron to the litre, and 2 grammes of carbonic acid. They are clear, without colour or smell, and can be kept without alteration in well-closed vessels.

2nd. *Mixed bicarbonated springs.*—This group, like the preceding, is composed of three principal springs, the *Source du Pavillon, Source Desaix,* and *Source des Pyramides.* The Pavillon spring is the typical one, being the richest in mineralised constituents, which are represented by soda, potash, magnesia, and lithia. Iron and chloride of sodium form but small elements in its composition.

3rd. *Bicarbonated magnesian springs.*—These resemble the Pavillon spring in their chemical composition; but, like the Morny spring, they add to the alkaline principles, purgative salts : 43 centigrammes of magnesia and 40 centigrammes of sulphate of soda to the litre, impart a special character to their therapeutical action.

The Establishments stand at the bottom of the valley ; they have been lately restored and re-organised, and possess several *buvettes,* a sufficient number of bath-rooms, and special apartments for douches and inhalations.

The complaints to which the waters of Châteauneuf are chiefly suited may be divided into two classes,— the first including rheumatism and the manifestations of the arthritic diathesis, and the second chloro-anæmic affections, and dyspepsia in its various forms.

The variety of temperature in these numerous springs allows of their application to very diverse cases, and of the graduation of the treatment according to the particular constitution of the patient.

The waters of Châteauneuf are exported, and keep without any deterioration.

CHATELDON

(*Puy-de-Dôme*)

CALCAREO-BICARBONATED GASEOUS SPRINGS (SLIGHTLY FERRUGINOUS)

CHATELDON is a small town situated about ten miles from Thiers, and not far from Vichy, in a valley surrounded by steep and rocky hills. Its mineral waters, which are calcareo-bicarbonated, proceed from five springs, the temperature of which varies between 9° and 13·6° C. (48° and 56° F.) Their mineralisation is similar in character, but not equal in quantities : it varies between 3·424 grammes and 5·128 grammes to the litre.

The waters are clear, they sparkle in the glass, and at the source they are constantly giving off bubbles of carbonic acid gas ; their taste is piquant, sharp, and slightly ferruginous.

The diseases in which they produce beneficial effects are certain affections of the urinary passages, calculous nephritis, gravel, vesical catarrh, dysuria, and retention of urine; various forms of dyspepsia, anorexia, accompanied with bad taste in the mouth, furred tongue, and all the train of symptoms connected with defective digestion. They have been recommended in those cutaneous cases which often coincide with latent functional derangement of the digestive organs, such as nettlerash, *couperose*, impetigo, &c.

The Chateldon waters are chiefly used for exportation. They are often drunk as table waters, but the iron which they contain renders their use, without medical advice, imprudent.

VICHY

(*Allier*)

GASEOUS SODIO-BICARBONATED, OR GASEOUS FERRUGINOUS BICARBONATED, HOT OR COLD SPRINGS.

VICHY is now supposed to have been the ancient *Vicus Calidus* of the Romans, since excavations have resulted in the discovery of undoubted traces of antique baths. It is a charming little town of 6000 inhabitants, built on the banks of the Allier, in the middle of a fertile basin surrounded by green hills of a moderate height. This long-famous Thermal Station is incontestably one of the most important in France, from the three points of view—of its waters, its Thermal Establishments, and the ever increasing number of its visitors. At present the latter exceeds 25,000 in the season.

The *Thermes de Vichy*, which are the property of the State, and are administered by a company, may claim to be among the finest in the world. They include two large principal Establishments, and a secondary one, called the *Bains de l'Hôpital*.

The first of these buildings, set apart for first-class baths, contains 100 bath-rooms, without counting apartments for douches of every variety. The façade, which looks upon the park, is pierced with seventeen imposing arches. An immense gallery, which serves as a covered *promenoir*, traverses the building from north to south, and gives access to the bath-room galleries, the windows of which look out on the gardens. The east gallery is reserved for ladies, and the west for men. In each are fitted up libraries and reading-rooms. At the end of the great promenade gallery are placed, to the right the carbonic acid baths and inhalation-rooms, to the left the offices for entering the names of patients and selling bath tickets.

The second building, appropriated to second and third-class baths, which are entirely separate, is of a rectangular form. It includes 180 bath-rooms of the former, and twenty-four of the latter class, without counting douche-rooms. Like the other building, this Establishment is traversed by a promenade gallery connecting the bath galleries, and, like the other also, one side is set apart for ladies, the other for men.

It is right to mention that the difference between the three classes of baths consists only in the luxury of the furniture, and the quantity of linen, &c., supplied to each person.

The *Bains de l'Hôpital*, thus called because they are opposite the spring of that name, were entirely rebuilt

in 1875. The façade is pleasing. The establishment
includes thirty-four bath-rooms, several douche-rooms,
and a public bath for ladies. It is supplied by the
Source de l'Hôpital, which (as well as the little square
where it rises) has been the centre of such alterations
that the aspect of that quarter of the town has been
completely changed.

Vichy also possesses a military Hospital, founded in
1846. It contains 120 rooms for officers, and dormi-
tories for sixty soldiers, who can take their baths
and douches in the Hospital itself, where a Thermal
Establishment has been specially constructed for
them.

By means of all these numerous and abundant conveniences, Vichy can give 3000 baths a day.

The springs of Vichy and its neighbourhood come
to the surface after traversing the tertiary calcareous
marls, and ancient alluvial deposits of the river Allier.
According to the engineer François, they seem connected by position and origin, either with the quartziferous rocks of red porphyry, of which the valley of
the Allier distinctly marks the western limit, from the
rising ground of Vichy to above Chateldon,—or with
the basalt and trap rocks which crop up through the
porphyry.

The Government springs which supply the great
Establishments are nine in number. The following
table gives their names, temperatures, and products in
twenty-four hours.

TABLE OF TEMPERATURE AND YIELD OF EACH SPRING
IN THE 24 HOURS

	TEMPERATURE (Centigrade)	LITRE (1000 gramm.)
Grande Grille . . .	42°	98·000
Puits Carré	45° ⎫	
Chomel	44° ⎭	200·000
Hôpital	30°	60·000
Célestins (old) . . .	14°	0·300
Célestins (new) . . .	14°	3·000
Lucas	29°	86·000
Brosson or du Parc . . .	22°	48·000
Mesdames	23°	20·000
Hauterive	14°	86·000

Besides the springs belonging to the State, there
are also at Vichy two others, which are the property
of private individuals. These are, the *Lardy* spring
(ferruginous), which supplies a very well-arranged
little Thermal Establishment, to which is added a very
complete system of hydropathic treatment; and the
Larbaud (alkaline), which is not connected with any
establishment.

The *Source de la Grande-Grille* is perhaps the best
known of all. It took its name from a large iron gate,
which used formerly to protect it. It is situated in
the north-east corner of the chief Establishment, at one
of the ends of the gallery of the springs. The *buvette*
and its appurtenances are installed in a little recess,
surrounded by a railing, and entered by a flight of two
steps on each side. For a long time this spring only
rose at one place, but now it comes out at two alter-
nately,—at one by day, and the other by night. By

day it flows at the *buvette,* and furnishes to it about 16,500 gallons of water ; by night it emerges at a lower point, which is not seen by the public, and gives about 21,000 gallons, which are used for the supply of the baths, and also for exportation.

The *Puits Carré* was formerly called *Source des Capucins.* It is situated in the middle of the north gallery of the Establishment, and can only be seen by visiting the subterranean vaults. For a long time it supplied a *buvette* which has now been removed, and at present all the water which it gives is employed for the service of the baths.

The *Source Chomel,* discovered in 1775 by Professor Chomel, emerges near the preceding spring, and unites its waters with those of the *Puits Carré* to supply the baths.

The *Source de l'Hôpital* rises in a vast circular stone basin, opposite the Civil Hospital, placed upon four flights of steps, and raised nearly six feet above the level of the ground. It supplies the *buvette* and the little Establishment called *Bains de l'Hôpital.*

The *Source des Célestins* owes its name to a convent of Celestins which formerly stood on the spot, and of which some ruins are still to be seen. The spring is situated behind Old Vichy, on the banks of the Allier, and rises immediately at the foot of an enormous rock of aragonite; a large square basin, hewn in the stone, receives the waters as they rise, and a pumping system brings them to the level of the soil. At a little distance from the spring there is a rustic summer-house, from which a covered passage leads to the *buvette;* while a little garden, laid out in the English fashion, surrounds the spring.

The *Nouvelle Source des Célestins* comes from the same rock as the preceding, and the *buvette* is in the same garden.

The *Source Lucas* is opposite the Military Hospital. Its water is drawn off by pumps, and divided between the *buvette*, the reservoirs of the Thermal Establishment and the baths of the Military Hospital.

The *Source Brosson* or *Source du Parc* in front of the Thermal Establishment, was obtained by a boring 158 feet in depth, and presents very irregular intermittences, usually lasting from forty-five to fifty-five minutes. Every eruption of the water is preceded by a considerable emission of gas, and accompanied by violent noises. The water then rises, and flows in sudden and irregular jets. It is very rich in carbonic acid, and supplies a *buvette* protected by an elegant pavilion; but it is chiefly employed for the service of the baths.

The ferruginous spring of *Mesdames* corresponds to that of *la Grande-Grille,* being situated at the opposite (north-west) corner of the Establishment. It is brought to this point by conduits, from its rise on the Cusset road, in the avenue called *Mesdames*, nearly a mile from Vichy.

The *Puits d'Hauterive* is a spring obtained by boring, in the village of Hauterive, about three miles from Vichy. For some time it supplied a little thermal establishment; but this is now suppressed, and the water is exclusively used for exportation.

The *Source Lardy* is situated within the ancient precincts of the Celestins convent. It rises from a boring nearly 500 feet deep, and yields daily more than 1500 gallons of water, at a temperature of 25° C. (77° F.) It contains per litre 9·165 grammes of mineralising elements, of which carbonic acid represents 1·75 gram., bicarbonate of soda 4·91 gram., bicarbonate of protoxide of iron 0·028 gram., and arseniate of soda 0·003 gram.

The *Source Larbaud* has a temperature of 22° C. (72° F.), and contains per litre 4·85 gram. of bicarbonate of soda. It is principally used for exportation.

All the Vichy waters, from whatever spring they proceed, resemble each other in appearance and taste, and differ only in their degree of thermality. They are pure, transparent, and gaseous. When taken up in a glass, they release a quantity of bubbles of carbonic acid gas, which settle on the side of the glass, or rise to the surface. These waters are sharp and piquant to the palate, but have a slightly nauseous flavour, and a dull smell.

When the chemical composition of the Vichy waters is attentively examined, bearing in mind the physiological action of the mineralising principles which they contain, it will be easily seen that they are first stimulating, then alterative, and finally tonic and restorative.

We now come to the therapeutical properties of

CHEMICAL ANALYSIS

WATER 1000 grammes (1 litre).	SOURCE DE LA GRANDE-GRILLE.	SOURCE DU PUITS CHOMEL.	SOURCE DU PUITS CARRÉ.	SOURCE LUCAS.	SOURCE DE L'HÔPITAL.	SOURCE DES CÉLESTINS.	SOURCE DU PUITS BROSSON.	SOURCE DU PUITS MESDAMES.	SOURCE D'HAUTE-RIVE.
Carbonic acid, free	Litre. 0·908	Litre. 0·768	Litre. 0·876	Litre. 1·751	Litre. 1·067	Litre. 1·049	Litre. 1·555	Litre. 1·908	Litres. 2·183
	Grammes.	Grammes.	Grammes.	Grammes.	Grammes.	Grammes.	Grammes.	Grammes.	Grammes.
Bicarbonate of soda	4·883	5·091	4·893	5·004	5·029	5·103	4·857	4·016	4·687
,, potash	0·352	0·371	0·378	0·282	0·440	0·315	0·292	0·189	0·189
,, magnesia	0·303	0·338	0·335	0·275	0·200	0·328	0·213	0·425	0·501
,, strontia	0·003	0·003	0·003	0·005	0·005	0·005	0·005	0·003	0·003
,, lime	0·434	0·427	0·421	0·545	0·570	0·462	0·614	0·604	0·432
,, and oxide of iron	0·004	0·004	0·004	0·004	0·004	0·004	0·004	0·026	0·017
Sulphate of soda	0·291	0·291	0·291	0·291	0·291	0·291	0·314	0·250	0·291
Phosphate of soda	0·130	0·073	0·028	0·070	0·046	0·091	0·140	Traces	0·046
Arseniate of soda	0·002	0·002	0·002	0·002	0·002	0·002	0·002	0·003	0·002
Chloride of sodium	0·534	0·534	0·534	0·518	0·518	0·534	0·550	0·355	0·534
Silicic acid	0·070	0·070	0·068	0·050	0·050	0·060	0·055	0·032	0·071
	7·914	7·959	7·833	8·797	8·222	8·244	8·601	7·811	8·956

Traces of bicarbonate of manganese, borate of soda, and bituminous organic matter.—*Bouquet.*

Vichy water. Looking alone to the customary practice at Vichy, and to the particular reputations of the springs, it might be supposed that each of these possessed special properties for the cure of one of the complaints treated at this Thermal Station. In fact, the *Source de l'Hôpital* seems to be reserved for liver complaints, the *Source des Célestins* for gout and affections of the urinary passages, the *Source de la Grande-Grille* for lymphatic persons, or those who have been greatly weakened by malarious fevers (paludal cachexia), &c. "This practice has no doubt some *raison d'être*," says Dr Durand-Fardel, the distinguished hydrologist, "but if one attached to it any idea of speciality properly so-called—of such a spring curing such a group of diseases—one would often be mistaken. In choosing the spring, one must habitually pay as much attention to the general condition of the patient as to the nature of his illness. Only, as the greater part of the patients attacked by the same disease are on the whole in a similar condition, the consequence is that for the majority among them the springs indicated will be the same."

The diseases treated at Vichy are very numerous; it is needless to say that they are all chronic. A few of the principal may be noted here.

Affections of the digestive organs.—These diseases are much the most numerous in the thermal practice of Vichy, and the efficacy of the waters is fully equal to their reputation, chiefly for dyspepsia with acidity, foul breath, and flatulence, chronic gastritis, enteritis, dysentery, and long-standing diarrhœa, with the localised congestions which so often accompany them. In these various cases, the water of the *Hôpital* spring, given in small doses, acts marvellously, restores appetite, facilitates digestion, soothes pain, and often speedily cures the diarrhœa. The waters of the *Célestins* and *Mesdames* springs agree better with some patients.

Liver complaints with or without calculus.—Vichy waters do wonders in the greater number of liver complaints. While rendering the bile more fluid, they stimulate the organic vitality of the hepatic cells, increase the activity of the circulation in the capillaries of the liver, and impart more elasticity to the whole organ ; they are also eminently tonic and resolvent. Their excellent effects may be above all relied upon against hypertrophy, without morbid products or degeneration.

In cases complicated with biliary calculi, there is no doubt that the Vichy waters have often favoured their ejection, by stimulating the contractility of the biliary ducts, or the gall bladder which serves as a reservoir. It would even appear that the waters would lead to their dissolution, if (as is believed to have been experimentally demonstrated) the formation of these calculi results from a want of alkalinity in the bile. In this case, the waters would act by supplying to it the alkaline matter in which it was deficient.

Abdominal congestions resulting from malarian fevers.—The enlargements of the spleen and mesenteric glands, which are of a paludal origin, sometimes yield in a quite unhoped-for manner to the well-directed use of these waters. In this respect, they can compete successfully with those of Carlsbad.

It is well known that Vichy receives every year from the French colonies a large number of soldiers and sailors, worn out by long years of hepatic and paludal complaints, which are endemic in hot countries. "In this case," says M. Daumas, physician to the Military Hospital, "it is marvellous to see with what facility livers which were enormously enlarged, passing beyond the umbilical region, and invading a large part of the abdominal cavity, seem—as it were—to melt away (after two or three weeks of treatment), under the eye of the observing physician; while at the same time there is improvement in the general symptoms, the digestive functions are restored to their normal conditions, the icteric hue of the skin disappears, and everything announces the revival of the vital energies and the return of health." It is needless to add that the internal use of the waters is combined with that of warm or cold mineral douches upon the ⁓diseased organ.

Gravel.—Vichy waters possess an incontestable efficacy against that particular form of gravel which is called *red* gravel or uric acid. Their action is sometimes so rapid that—after the first few glasses— patients perceiving that no more gravel is excreted are alarmed, thinking that it remains lodged in the internal organs; while, on the contrary, it has been dissolved. This dissolution may be explained by saying that the excess of uric acid has combined with the soda of the water to form a urate of soda, which has passed away with the urine. Sometimes, however, the Vichy water acts less as a chemical agent than as a stimulant to the renal functions, and in this case the gravel, instead of being dissolved, is simply expelled from the kidneys, and carried away by the urine.

When *graviers* or small calculi are present, they are expelled rather at the end than at the beginning of the course of treatment, since a certain time is required before they detach themselves.

Diabetes.—" The Vichy waters," says Dr Durand-Fardel, " offer to the diabetic only a palliative treatment; but it is one which affects the most essential characteristics of the malady, and which in a very high degree assists the imperfect resources of hygiene and therapeutics. This action of the waters, incomplete as it is, yet nevertheless is sufficiently distinct for their use to be placed among the remedies most constantly indicated in the treatment of diabetes."

Gout falls within the range of the Vichy waters, but here a specific curative action on their part must not be expected. They merely attenuate its manifestations. Under their influence, attacks of acute and periodic gout become less frequent and less severe; the deformities and stiffness resulting from chronic gout are modified and diminished, and limbs which were perfectly powerless have been seen to recover most of their motions. But gouty patients ought to be prudent, and take the water in small doses, only in the intervals between the attacks, and at the longest possible time from the latest.

Uterine affections.—The baths, assisted by the irrigations, for which every ladies' bath room is provided with an apparatus, easily conquer chronic congestions and other affections of the uterus.

Skin diseases.—There are certain diseases of the skin which are exasperated by sulphurous waters, and which, on the contrary, do well at the springs of Vichy. It is probable that the alkaline waters act here in a double manner; on the one hand, locally, against the alteration of the derma; and on the other, as a regulating influence, when there is functional disturbance.

When the Vichy waters travel, they are exposed to two causes of alteration, viz., oxidisation, and loss of carbonic acid gas. The various springs preserve their integrity of composition and properties, in degrees varying in the following order :—(1) *Hauterive ;* (2) *Célestins ;* (3) *Grande-Grille.* Ferruginous springs : (1) *Lardy ;* (2) *Mesdames.*

Their exportation increases every year, and the numbers have now risen to three-and-a-half million bottles. This partly results from the fact that almost all patients who have spent a season at Vichy, continue to use the water after their return home.

For some years, the Company which farms the springs has also been exporting the salts extracted from the waters in the form of rolls, intended for preparing baths of artificial Vichy water. In order that fraud may be impossible, the export both of the water and the salts is placed under Government control, by a ministerial decree dated as far back as 1857. It is clear that Vichy waters taken at home under these various forms, will never replace a season at the Station itself, but they will always be useful, and may be beneficially employed in all cases where they are therapeutically indicated.

The thermal season at Vichy is from April 15th to September 15th; but as the great heat of July and August seems to defeat the good effects of the waters upon many patients, the months of May and June should rather be recommended, as well as the latter half of August and the following fortnight of September.

Vichy possesses a very fine casino, which is in no way inferior to the finest establishments of its kind. It contains reading-rooms, conversation-rooms, ball-rooms, and others for concerts, billiards, and cards, a ladies' drawing-room, and a theatre. Very fine grounds surround it on all sides. The most frequented promenade is the *Allée des Dames*, which is bordered with superb poplars, and runs along the picturesque banks of the Sichon.

Among the principal excursions in the neighbourhood, visitors may be especially directed to *Cusset*, the *Montagne Verte*, the Valley of *Malavaux*, the hill of *Saint-Amand*, the *Ardoisière*, the ruins of the château of *Montgilbert*, the *Grotte des Fées*, the châteaux of *Charmeil* and *Randan*, &c.

CUSSET

Cusset is a very ancient and interesting little town, close to Vichy,—in fact, they are not two miles apart. It possesses a Thermal Establishment, very well managed, though constructed on a small scale. It contains twenty-four bath rooms, six douche rooms, a small room for ascending douches, one hall for mineral vapour baths and douches, and one for general and local applications of carbonic acid gas. It is supplied by three springs, at a temperature of 16° C. (61° F.)

This establishment, which has the misfortune of being too close to Vichy, is only of local utility, and would not suit strangers.

SAINT-GALMIER

(*Loire*)

CALCAREO-BICARBONATED, GASEOUS, COLD SPRINGS.

Saint-Galmier is a small town of 3000 inhabitants, about twelve miles from *Montbrison*, on the slope of a hill below which flows the river *Corse*.

Its mineral waters, which are now universally known, and are exported by millions of bottles annually, proceed from four springs. They are all cold bicarbonated, mixed (containing soda, lime, and magnesia), and are almost identical in composition.

CHEMICAL ANALYSIS

WATER 1000 grammes (1 litre).			SOURCE RÉMY.
Carbonic acid, free . . . 1½ vol.			
			Grammes.
Bicarbonate of lime			0·780
„ soda			0·089
Sulphate of magnesia			0·741
Chloride of sodium			0·200
Alumina and oxide of iron . . .			0·020
Insoluble residue			0·020
Bouis.			1·850

The waters of Saint-Galmier are clear and inodorous, sparkling, of a very pleasant and piquant taste, and giving off a great quantity of carbonic acid gas. They are only used as table-waters, and being slightly stimulant, aperient, digestive, and diuretic, are suitable to persons of slow and difficult digestion, suffering from dyspepsia or gastralgia.

RENAISON

(*Loire*)

CALCAREO-BICARBONATED GASEOUS, COLD SPRINGS.

RENAISON is a large village about six miles distant from Saint Galmier, in the oval basin of the Loire. Its waters are cold, bicarbonated, mixed (containing soda, lime, and magnesia), gaseous, and present a strong analogy with those of Saint-Galmier.

There is no Thermal Establishment at Renaison. Its waters are chiefly exported. They are very agreeable to the taste, and may advantageously replace the artificial seltzer water, which has been so greatly misused.

SAIL-SOUS-COUZAN

(*Loire*)

MIXED BICARBONATED AND GASEOUS SPRINGS

COUZAN, or *Sail-sous-Couzan,* is a village in the district of Montbrison, situated at the foot of a conical mountain, crowned by the vast and imposing ruins of the castle of Couzan.

A few years ago, Couzan was quite unknown as a Thermal Station; it had then only one spring, and has now two.

The *Source Fontfort* gives about 2208 gallons per day, and the *Source Rimaud* 6174 gallons. These two springs supply a recently-constructed Establishment, which comprises twenty-six baths, each provided with an aspiration tube for carbonic acid, douches of all sorts, vapour baths, and, in short, all the material for modern hydropathic treatment.

The Couzan waters are sodio-bicarbonated and ferruginous, acidulated, somewhat sharp and pleasant to the taste; they sparkle and bubble vehemently, like champagne. They leave on the soil a considerable

deposit of hydrated oxide of iron, and gradually grow
thick under exposure to the air.

CHEMICAL ANALYSIS

WATER 1000 grammes (1 litre).				SOURCE RIMAUD.
				Grammes.
Carbonic acid, free	.	.	.	0·4317
Bicarbonate of soda	.	.	.	1·9509
,, potash	.	.	.	0·3034
,, lime	.	.	.	0·3870
,, magnesia .	.	.	.	0·3436
,, iron	.	.	.	0·0177
Chloride of sodium	.	.	.	0·0876
Silica .	.	.	.	0·0410
Sulphate of lime	.	.	.	0·0465
Traces of bicarbonates of manganese and lithia, iodide of sodium, arseniate of soda, alumina and organic matter.				3·6094

These waters are exciting at the commencement of
the cure, although in most cases they agree well with
the stomach. The baths at first produce restlessness and
insomnia, but these unimportant discomforts are soon
followed by a period of sedative and restorative influences. They are recommended for internal use, for baths
of moderate temperature, and cold douches, as a remedy
for the long-continued dyspepsia which is frequently
to be found in anæmic patients and convalescents ; and
also against dyspepsia when supervening in different
forms of cachectic diseases, and when connected (as
it so often is) with uterine affections. As to the
variety of the dyspepsia itself—whether it be atonic,
flatulent, bilious, or accompanied with acidity,—
Couzan water is equally efficacious in combating it.

Some diseases of the uterus (congestion, enlargement, catarrh) are successfully treated by the carbonic acid douche, either pure or mixed with water.

The neighbourhood of Couzan offers to patients numerous and attractive objects for walks :—The ruins of the castle of *Couzan,* the abbey of *Leigneux,* the cascade of *Gourmé,* the grottos of the *Fayettes,* the hill of *Pierre-sur-autre,* &c.

SAINT-ALBAN

(*Loire*)

FERRUGINOUS, BICARBONATED, AND GASEOUS, COLD SPRINGS.

SAINT-ALBAN is a hamlet situated nearly five miles from *Roanne,* on the left bank of the Loire, at the foot of the range which terminates the series of the mountains of *Forez.*

This watering place, which has been very prosperous for several years, possesses four mixed bicarbonated springs, called the *Puits de César, Puits d'Antonin, Puits Neuf,* and *Puits de Faustine,* which supply a very fine Thermal Establishment, including numerous bath-rooms, a complete set of douches, a large public bath, and two inhalation rooms. Their waters are clear and colourless, of a piquant smell and flavour, with a ferruginous after-taste; they allow numerous bubbles of carbonic acid gas to escape, and deposit at

the bottom of the reservoirs a reddish sediment.
The chemical composition is almost the same in all.

CHEMICAL ANALYSIS

WATER 1000 grammes (1 litre).	SOURCE DU PUITS DE CÉSAR.
	Grammes.
Carbonic acid, free	1·9499
Bicarbonate of soda	0·8561
＂ potash	0·0834
＂ lime	0·9382
＂ magnesia	0·4577
＂ iron	0·0233
Chloride of sodium	0·0301
Silica	0·0451
Traces of iodide of sodium, arseniate of soda and organic matter.—*Lefort*.	4·3838

These waters are stimulating, diuretic, digestive,
tonic, and restorative. They are used in baths and
douches, but chiefly for drinking. The carbonic acid
gas which they contain in considerable quantity is
utilised for inhalation, as at Couzan.

The therapeutical applications of Saint-Alban water
are the same as those of Couzan. They will therefore
be prescribed in the different varieties of dyspepsia,
and chiefly in that which results from atony of the
digestive organs; in the gastralgia which so often
accompanies chlorosis and anæmia; in sub-acute
nephritis and catarrhs of the bladder, and in chronic
affections of the uterus. Inhalations of carbonic acid
have been successfully used in cases of spasmodic
asthma, chronic pharyngitis, and asthenic aphony.

The employment of carbonic acid gas as an internal

and external remedial agent gives an immense advantage to this Station, and has, in fact, created for it a sort of speciality. Patients are no longer obliged to go to Germany to obtain this kind of treatment, but can find at Saint-Alban carbonic acid baths, douches, and inhalations; and both the apparata and all the details of the treatment are more complete and better organised than can be found anywhere else.

The season lasts from June 1st to September 15th.

The water is exported in large quantities; but it is not a simple table-water, and ought not to be taken except by medical advice.

Among the places of interest in the neighbourhood of Saint-Alban may be noted the *Vallée du Désert*, the village of *Embierle*, the ruins of the châteaux of *Saint-Georges* and *Jacques-Cœur*, &c.

NÉRIS

(*Allier*)

SODIO-CHLORINATED AND BICARBONATED WARM SPRINGS

To judge by the débris of its Circus and the ruins of its ancient Thermæ, once adorned with numerous columns, the remains of which have been discovered, there can be no doubt that Néris—now only a small town with a population of 2400—was formerly a rich and handsome city, at the period when the Romans ruled in Gaul. It stands in the upper portion of a valley through which flows an affluent of the river Cher, and is divided into two parts, the upper and lower town.

There are two Thermal Establishments, the *large* and the *small*. There is also a hospital, where the poorer patients can live.

The large Establishment is indisputably one of the finest and most complete, either in France or abroad. It includes four *piscines*, two of which, extremely large and of moderate temperature (32° to 34° C., or 90° to 93° F.), serve as swimming-baths; while the two others, which are hot (38° to 42° C., or 100° to 107° F.), and of lesser dimensions, are used for partial baths of short duration. There are sixty-two private baths, placed in an equal number of small rooms, and fitted with descending douches of every variety of temperature. There are also special douches in apartments provided for the purpose; besides large rooms for vapour baths, *massage*, and inhalations. To all this must be added complete arrangements for hydropathic treatment.

The small Establishment was reconstructed a few years ago; it contains four *piscines*, several douche rooms and vapour-baths, and is exclusively reserved for the indigent patients residing in the Hospital.

These Establishments are supplied by six springs, known under the names of *Puits de la Croix, Puits de César, Puits Carré, Puits Boirot, Puits du Noyer,* and *Puits Falvard.* They contain chloride of sodium and bicarbonates, and appear to proceed from the same underground sheet of water. The total daily product is over 242,000 gallons of a clear water, colourless in the glass, but appearing green in the wells, in consequence of the confervæ which it deposits on the sides and bottom; it is inodorous, and of a slightly saline taste. The temperature varies between 46° and 53° C. (115° and 127° F.) . The spring called *Puits de la Croix* supplies the *buvette*.

One important matter to be noticed is the great quantity of free gas emitted by these springs, in proportions as under :

Nitrogen (azote)	88·62
Carbonic acid	11·48
	100·00

CHEMICAL ANALYSIS

WATER 1000 grammes (1 litre).	SOURCE PUITS DE CÉSAR.	SOURCE PUITS DE LA CROIX.
	Cubic centimètres.	Cubic centimètres.
Carbonic acid, free . .	0·049	0·039
Nitrogen . . .	13·000	10·200
Oxygen . . .	—	1·100
	Grammes.	Grammes.
Bicarbonate of soda . .	0·4169	0·4167
,, potash . .	0·0129	0·0125
,, magnesia . .	0·0057	0·0057
,, lime . .	0·1455	0·1463
,, iron . .	0·0042	0·0033
Sulphate of soda . . .	0·3896	0·3848
Chloride of sodium . .	0·1788	0·1782
Silica	0·1121	0·1030
	1·2657	1·2505

Traces of bicarbonate of manganese, iodide of sodium, fluoride of sodium, and nitrogenous organic matter.—*Lefort.*

The waters of Néris are chiefly used externally, although some patients drink them. They have a sedative action upon the nervous system, but are at first exciting.

Among the complaints treated at Néris, the first place must be given to the diseases of the nervous system, which are its true speciality. Physicians, therefore, send to this station patients suffering from neuralgia, sciatica, convulsions, St Vitus' dance, paralysis, nervous tremblings, chronic vomiting, hysteria, and certain forms of sleeplessness. After these come rheumatic and gouty affections, white swellings, chronic metritis connected with a neurotic condition; and, lastly, cutaneous affections of a herpetic nature, such as lichen, eczema, impetigo, acne, &c.

The number of visitors to Néris has lately increased in considerable proportion; for its waters best replace —for medicinal purposes—those of Baden-Baden and Teplitz, to which the chief attraction was the prospect of amusement. Life at Néris, on the contrary, is very quiet. This results from the class of invalids who frequent this Station,—nervous patients who seek for silence, solitude, and quietness. The only recreations to be procured are reading, conversation, and delightful walks in the surrounding country. It may be added that life is easy and inexpensive, and that there are very good hotels.

BOURBON-L'ARCHAMBAULT

(*Allier*)

SODIO-CHLORINATED WARM SPRINGS, WITH IODIDES AND BROMIDES.

BOURBON-L'ARCHAMBAULT is a small town of 3000 inhabitants, situated about fourteen miles from *Moulins*, in a healthy valley, watered by the river *Bruge*, and surrounded by high hills.

The Thermal Station has been one of the most celebrated in France. Catherine de Médicis, the Princes of Condé, Racine, Boileau, Turenne, Mme. de Sévigné, and Mme. de Maintenon, all visited it in their turn. It was here also that Mme. de Montespan, after her fall from favour, passed the latter part of her life in penitence and religious practices.

The Thermal Establishment contains eight small *piscines*, of a square shape, built of cut stone, and each furnished with a douche. The northern one, which is larger than the others, is still called the *cabinet du prince*, in memory of the famous diplomatist Talleyrand, who bathed in it for thirty years in succession. There are also a large number of bath-rooms on the ground floor and first floor, besides apartments for douches and hot-air baths. Rooms are also arranged for the pulverisation of the mineral water, and eye-douches. Patients can be carried from their beds to the baths and back again, in closed chaises-à-porteurs.

Bourbon-l'Archambault also possesses a Civil and a Military Hospital, containing 300 beds, appropriated to poor patients, and to soldiers.

The springs are two in number; the *Source Chaude* and *Source Jonas*. The first is the most important; it is sodio-chlorinated, and its temperature is as high as 52° C. (126° F.) It yields daily over 462,000 gallons of a water which is at first colourless and inodorous, but afterwards grows thick, becomes covered as it cools with a calcareous pellicle, and gives off a slight smell of sulphuretted hydrogen. Owing to the escape of gas, the spring is constantly bubbling. Both springs deposit confervæ on the sides of their basins, but those at the *Source Chaude* are observed to be a shade darker in colour.

CHEMICAL ANALYSIS

WATER 1000 grammes (1 litre).	SOURCE CHAUDE.
	Litre.
Carbonic acid, free . . .	0·166
	Grammes.
Bicarbonate of lime	0·507
„ magnesia	0·470
„ soda	0·367
Sulphate of lime	0·220
„ potash	0·011
Chloride of calcium	0·070
„ sodium	2·240
Alkaline bromide . . .	0·025
Silicate of lime and alumina . .	0·370
„ soda	0·060
Crenated oxide of iron . . .	0·017
Traces of chloride of potassium and organic matter.—*O. Henry.*	4·357

As appears from the above analysis, this water contains chloride of sodium, with iodides and bromides. Taken internally, it is eminently diaphoretic, diuretic, tonic, and restorative. In hot baths at from 34° to 38° C. (93° to 100° F.), it is highly exciting to the circulation and the nervous system, and consequently it is only used at this temperature with great caution. In warm baths at from 28° to 31° C. (82° to 88° F.), it gently stimulates the skin, and has a sedative action. That of the *Jonas* spring is diuretic and laxative.

These waters are celebrated for their efficacy against scrofula and *lymphatism*, arthritis, adenitis, local congestions in the cellular tissue, ulcers, and fistulous openings. Most diseases of the bones are modified or cured by these waters, and the constitution of the patients is, at the same time, modified and regenerated. Ophthalmia arising from the same cause, blepharitis, keratitis, &c., are treated most successfully with douches and instillations administered drop by drop by a special apparatus, and according to a particular system.

Rheumatism, as well as scrofula, is a speciality of this station. All forms of chronic rheumatism are here advantageously treated, and often completely cured. Rheumatic patients of lymphatic constitutions obtain the greatest benefit from these waters.

Paralysis is another affection in which very good results are obtained at Bourbon-l'Archambault. The local physicians are unanimous in testifying to the success of the waters upon cases of rheumatic or hysterical paralysis.

The town itself offers nothing very remarkable to the visitor. A pleasant walk may be had in the fine avenue of chestnut trees, planted by Madame de Montespan at a short distance from the ruins of an old feudal castle, standing on the shores of a lake three miles in circumference. The casino pavilion offers the usual variety of amusements.

SAINT-PARDOUX

(*Allier*)

FERRUGINOUS, BICARBONATED, COLD SPRING (FEEBLY MINERALISED).

SAINT-PARDOUX is a very small Station, situated about six miles from *Bourbon-L'Archambault.* The mineral water proceeds from a single spring, the produce of which exceeds 1100 gallons per day; it is cold, bicarbonated, and ferruginous, and has the closest analogy with those of Saint-Alban and Saint-Galmier. It is considered an excellent table-water, and is exported in large quantities. There is no Establishment.

Saint-Pardoux water is diuretic, aperient, and tonic, and is indicated in cases of dyspepsia, anæmia, atonic affections of the urinary organs, and general debility.

BOURBON-LANCY

(*Saône-et-Loire*)

SODIO-CHLORINATED HOT SPRINGS (FEEBLY MINERALISED)

BOURBON-LANCY is a small town of 4000 inhabitants, built on a high hill near the banks of the river *Lorry*. Its reputation dates from the first half of the six-teenth century, when Catherine de Médicis, wife of Henry II, being still childless after many years of marriage, was sent to Bourbon-Lancy by her physician Fernel. She took the waters in drinks, baths, and douches, and at the end of nine months she brought into the world an infant, who subsequently became Henry III.

This Station possesses a fine Establishment and two Hospitals. The former contains a capacious *piscine,* the large dimensions of which (fifty-nine feet by twenty-nine) render it really a swimming bath; some thirty bath-rooms, besides douche rooms and a *buvette.* The old Hospital, founded at the end of the seventeenth century, has its *piscines* and its special douches. The new Hospital is a magnificent building, whose exterior aspect and imposing proportions would never lead to a suspicion of its destination; it was founded by the Marquis d'Aligre, and contains 400 beds.

The waters of Bourbon-Lancy are strongly impreg-nated with chloride of sodium. They proceed from seven springs, of which six are hot and one cold.

They are all brought together in the courtyard of the Establishment, after having risen at the base of the same perpendicularly cut rock in the Faubourg Saint-Léger. This proximity at their sources gives rise to the supposition that all the springs have a common origin, in spite of the differences in their temperatures. They yield altogether over 76,000 gallons per day of a clear, colourless water, without smell, and of a slightly saline taste, which gives rise in the basins to green confervæ, and emits bubbles of gas, that rise, and burst noisily at the surface. The quantity of carbonic acid gas to the litre has been stated at 0·034 gramme.

The analysis of the *Source d'Escure*, which is the most highly mineralised, may be taken as typical. The temperature of this spring in 54° C. (129° F.)

CHEMICAL ANALYSIS

WATER 1000 grammes (1 litre).	SOURCE D'ESCURE.
	Grammes.
Chloride of sodium	1·30
„ calcium	0·05
„ magnesium	0·40
Sulphate of soda	0·25
„ lime	0·02
Carbonate of lime	0·06
„ magnesia	0·15
Silica	0·02
Oxide of iron	0·02
Traces of iodide of sodium and arsenic.—*Tellier and Laporte.*	2·27

Taken internally, the waters of Bourbon-Lancy act on the mucous membranes, the secretions of which they increase, especially those of the digestive passages. In large doses they are purgative, but nevertheless have a tonic and restorative action, derived from their combined composition of chloride of sodium and iron. They stimulate the menstrual discharge, as well as the general circulation, and have a tendency to induce hæmorrhoidal congestions. They are very efficacious against all forms of rheumatism characterised by stiffness and enlargements of the joints; gout; different forms of paralysis; facial and sciatic neuralgia, and, in general, against all nervous affections. They act also on the heart, and have a beneficial effect upon functional disorders of that organ. Dyspepsia, flatulent gastralgia, chloro-anæmia, and various uterine complaints, are ameliorated by their use. And, lastly, they are successfully employed against scrofula, even in its gravest forms, such as arthritis, caries, and necrosis. Under their influence, wounds heal quickly, even those recently received in action, or produced by any other cause, as well as functional impediments resulting from old wounds.

Visitors to Bourbon-Lancy will not fail to notice in the Hospital the silver life-size statue of Madame d'Aligre, wife of the founder of that institution, to which he left at his death in 1848 a sum of three million of francs. Among the walks in the neighbourhood may be mentioned those to the ruins of the *Château Fort*, the small mountain of *Champaudé*, from which can be seen the Morvan Mountains, the forest of *Mont*, &c.

SAINT-HONORÉ

(*Nièvre*)

SODIO-SULPHURETTED WARM SPRINGS

SAINT-HONORÉ is a town situated about six miles from *Moulins-Engilbert,* among the wooded mountains of Morvan, nearly 1000 feet above the level of the sea. The mineral waters were known to the Romans, as is proved by the remains of the ancient Thermæ, brought to light by the excavations made in 1854, for the construction of the new Establishment. This is one of the most complete, and contains sixteen bath-rooms, with baths of stone and Nevers china, each provided with an ascending douche; three *piscines* with douches, furnished with all the apparata used in thermal hydro-therapeutics; several rooms for hip-baths, a very large public swimming-bath, in which the water is constantly renewed; a room for inhalation and pulverisation, and, lastly, five *buvettes.*

The exceptional geographical position of Saint-Honoré imparts to it a peculiar interest, since it is the only sulphureous Thermal Station in the central part of France.

The waters are sodio-sulphuretted and warm; they proceed from five springs, known under the names respectively of the *Source de la Crevasse, Source de l'Acacia, Source de la Marquise, Source Romaine,* and *Source de la Grotte.* They give altogether about

220,000 gallons per day of a clear water, perfectly transparent, of a slightly blueish hue, oily to the touch, with a sulphurous smell, and a bitter taste.

They are drunk, and used for baths, douches, inhalations, and pulverisations. When taken internally, they stimulate the appetite, and at the commencement of the treatment are slightly laxative; after a few days they are very well borne by the stomach. In baths, the water is oily, and covers the body with a multitude of gas bubbles which produce a highly stimulating effect upon the skin.

CHEMICAL ANALYSIS

WATER 1000 grammes (1 litre).	
	Cubic cent.
Sulphuretted hydrogen, free . . .	0·70
Carbonic acid, free . . .	$\frac{1}{9}$ vol.
	Grammes.
Carbonate of lime ⎫ „ maguesia ⎭ . . .	0·098
„ soda and of potash . . .	0·040
Silicate of potash ⎫ „ soda ⎭ . . .	0·034
„ alumina . . .	0·023
Alkaline sulphides . . .	0·003
Sulphate of soda . . .	0·132
„ lime . . .	0·032
Chloride of sodium . . .	0·300
„ potassium . . .	0·005
Oxide of iron and organic matter . . .	0·007
Traces of alkaline iodides, lithia and manganese.— O. Henry.	0·674

It appears from this analysis that the Saint-Honoré waters have a certain analogy of chemical composition with those of Eaux-Bonnes. The resemblance does not end there ; for, like the latter, they slightly stimulate the nervous and circulatory systems, excite and modify the secretions of the skin and of the mucous membranes ; increase the appetite, and facilitate digestion.

In the first rank of the diseases treated at Saint-Honoré must be placed affections of the respiratory passages, *i.e.* chronic laryngitis, chronic bronchial catarrh, and humid asthma. Pulmonary phthisis, in the first and second stages, is also here ameliorated by the inhalation treatment. Under its soothing and restorative influence, pulmonary congestion disappears, hæmoptysis is obviated, strength increases, and appetite improves.

After diseases of the respiratory passages, those which are most frequently to be met with at Saint-Honoré are *lymphatism*, scrofula, and their various manifestations ; amenorrhœa, dysmenorrhœa, leucorrhœa, uterine enlargements, and the nervous symptoms which often result from them.

Speaking generally, it may be said that these waters are also indicated in the treatment of all sero-purulent affections of the skin ; while dry or herpetic affections are either acted upon unfavorably or not at all.

To sum up, Saint-Honoré is a very important Station, in respect of the efficacy of its waters and its situation in the central part of France ; and it receives an annually increasing number of patients, who come to seek the amelioration or removal of their sufferings.

Life at Saint-Honoré is very agreeable, in conse-

quence of the numerous amusements provided in the
Establishment, and the charming walks in the neigh-
bourhood. Among the objects of the latter are : the
Vieille Montagne, the large and beautiful château
d'Espeuilles ; the ruins of the old château of *Glux-en-
Glaine*, near which is the source of the River Yonne ;
the marble quarries of *Champrobert*, the iron-works of
Fourchambault, &c.

POUGUES

(*Nièvre*)

COLD CALCAREO-BICARBONATED SPRINGS

POUGUES is a small town of 1350 inhabitants, about
eight miles from *Nevers*, in the valley of the Loire.
It took its place among Thermal Stations towards
the end of the fourteenth century, and for 300 years
was very much in fashion, owing to the patronage of
the different sovereigns and high personages who
came to seek from its waters the restoration of their
health. Among these illustrious visitors were Henry
II, Catherine de Medicis, Louis XIII, Louis XIV, the
Prince of Conti, &c.

After the end of the seventeenth century, Pougues
fell almost into obscurity; and it is only within the
last twenty years that it has regained its old celebrity
as a Thermal Station. Pougues owes its rehabilitation
to the numerous writings of Dr Felix Roubaud, who

has made known the true physiological and therapeutical properties of its waters.

These waters, which are cold calcareo-bicarbonated, proceed from two springs, known under the names of the *Source Saint-Léger* and *Source Bert*. The first, which has been the longest known, supplies a very comfortable hydro-mineral Establishment, to which it furnishes more than 11,000 gallons of water per day. The second has been lately discovered, and is exclusively used for drinking. It is weaker than the *Saint-Léger*, and is rather an' excellent table-water, than a medicinal agent. The water of both springs is exported in very large quantities.

The Establishment is composed of a main building with two wings, one reserved for ladies, the other for gentlemen. It contains twenty-six baths, very complete douche arrangements, and everything necessary for hydrotherapeutical treatment.

CHEMICAL ANALYSIS

WATER 1000 grammes (1 litre).	SOURCE SAINT-LÉGER.
	Grammes.
Carbonic acid, free	0·6091
„ of the bicarbonates	1·0098
„ of the carbonates	1·0033
Hydrochloric acid	0·1275
Sulphuric acid	0·1450
Silica	0·0150
Peroxide of iron	0·0146
Lime	0·7000
Magnesia	0·1150
Potash	0·0450
Soda	0·6290
Rivot.	4·4133

The waters of Pougues are employed with the greatest success in cases of dyspepsia which are of a purely nervous nature, and not symptomatic of either the herpetic or rheumatic diathesis. In their quality of alkaline waters, they act in a very remarkable manner against all the manifestations of the uric acid diathesis, gout, gravel, diabetes, and certain forms of albuminuria, which (according to Dr Roubaud) is the ultimate manifestation of the uric acid diathesis. They have also a very marked action on chronic nephritis and vesical catarrh.

The carbonates of lime and iron which they contain render them valuable in the reconstruction of the organism, in cases of chlorosis and anæmia, and those leucorrhœic and dysmenorrhœic symptoms, which in women so often accompany these diseases.

The basis of the treatment is the internal use of the waters, as is always the case with alkaline springs; their external use (baths and douches) may be considered as only supplementary to the former.

A handsome building has been erected opposite the park, containing a well-furnished reading-room, an elegant public sitting-room, and a very fine ball-room. On the whole, it may be said that life at Pougues is far from being dull.

Pougues water is exported, after having been charged with an additional quantity of carbonic acid gas, during the process of bottling.

PYRENEAN DIVISION

THE mineral waters of the Pyrenees were for the most part known to the Romans, who left there—as everywhere else—magnificent memorials of their presence. The springs are almost all characterised by the sulphureous element, so that they have a general superficial similarity, but the difference of their therapeutical properties creates (so to speak) a particular speciality for each.

The bathing arrangements of the Pyrenean Stations are perhaps more generally perfect than those in any other region; and when to this is added the extraordinary beauty of the mountains among which they are situated, it may fairly be said that this is the most favoured division of France.

EAUX-BONNES

(*Basses-Pyrénées*)

SODIO-SULPHURETTED SPRINGS, WARM AND COLD.

THE village of *Eaux-Bonnes* is situated 2460 feet above the level of the sea, in the wild Valley of Ossau, surrounded on all sides by lofty mountains, with bare summits, and slopes clothed in dark pine woods. It consists of a single street of houses, many storeys high, but scarcely harmonising in style with the tone of the surrounding landscape.

This Thermal Station, at present one of the most frequented in the Pyrenees, appears to have been known to the Romans.

There are at Eaux-Bonnes seven sodio-sulphuretted springs; the *Source Vieille*, the *Source Nouvelle*, the *Source d'En-bas*, the *Source d'Ortech*, the *Source Froide*, and two others recently discovered, which proceed from the same subterranean store of water as the five already mentioned. The two springs of recent discovery are exclusively appropriated to the service of the baths. The first three supply the Grand Establishment. The fourth rises on the left bank of the river Valentin, at the foot of the northern slope of the mountain, and the fifth springs up at the entrance of the ravine of La Soule. The *Source Vieille*, the longest and best known of the springs, is reserved for drinking purposes alone; it yields about ninety gallons per hour. The *Source d'En-bas* and the *Source Nouvelle* feed the baths. The *Source Froide*, like the *Source Vieille*, is employed for drinking.

The temperature of the springs varies from 12° C. (53·6° F.) to 33° C. (91·4° F.)

When it issues from the ground, the water of the springs is clear, and oily to the touch, with a strong smell resembling that of rotten eggs. Its flavour is sweetish, and not at all unpleasant; it scarcely leaves

behind it any hepatic taste, and patients therefore
drink it without any repugnance.

CHEMICAL ANALYSIS

WATER 1000 grammes (1 litre).				SOURCE VIEILLE.
			•	Grammes.
Sulphide of sodium	•	•	•	• 0·0210
Sulphate of lime	•	•	•	• 0·1750
Chloride of sodium	•	•	•	• 0·2640
Silicate of soda .	•	•	•	• 0·0310
Silica .	•	•	•	• 0·0320
Organic matter .	•	•	•	• 0·0480
Traces of sulphide of calcium, sulphates of potash, soda and magnesia, borate of soda, iodine and sulphide of iron.—*Filhol*.				0·5710

The Thermal Establishment, which has been recently
enlarged, is built in a simple and tasteful style. It
contains eleven private bath-rooms with marble baths,
a large hall for pulverisation, douches for the throat,
and two special rooms for *pédiluves,* the use of which
has lately increased to an astonishing extent. There
is also a *buvette,* situated against the side of the rock
called *Butte-du-Trésor.*

No better idea can be given of the physiological
action of the Eaux-Bonnes than by quoting the following passage, borrowed from a work by Dr Andrieux,
which has already stood the test of time, and bears on
each page the stamp of remarkable experience and
ability.

" General strength and activity are augmented,
sleep becomes unquiet, and the intellect is more active.
The beats of the heart increase in number and strength,

the pulse is fuller, more frequent, and more vigorous; the catamenial and hæmorrhoidal fluxes become more abundant, appear for the first time, or are re-established if they have been previously suppressed. The hæmorrhagic movement is directed from the centre towards the surface, blood escapes from the nostrils, the bronchial tubes, &c.; the appetite improves, and the muscular system of the intestines is stimulated into action, and its contractile power increased. The two great continuous systems of the vital economy, in which life is more specially concentrated, the nervous and the circulatory, have evidently undergone, in the forces which animate them, a modification which manifests itself in an exaggeration of their normal activity. The secretions in their turn are considerably modified; the cutaneous exhalation increases, as well as the urinary secretions. The mucous membranes become congested and reddened, and leucorrhœa increases; the nasal, laryngeal, and bronchial catarrhs are for the time intensified; expectoration becomes more abundant; pathological secretions of the skin are created, re-established, or exaggerated: but ultimately everything becomes regular, and the only feeling which remains is one of comfort."

Amongst the diseases of which the successful treatment has made the therapeutical reputation of Eaux-Bonnes, those of the respiratory organs must be given the first place. The action of the waters on the mucous membrane of these organs seems to be a true elective affinity, a fact which was first noted by Bordeu, and has since been confirmed by all his successors. According to M. Brian (the distinguished librarian of the Academy of Medicine, who for at least twenty years has practised at Eaux-Bonnes during the season), this action is chiefly felt in two different ways. First, it dissolves congestions and indurations, whether arising spontaneously or in consequence of anterior diseases of the bronchial tubes, the pleuræ, or the pulmonary tissues; secondly, it acts in a substitutive manner on apyretic inflammations of those organs, either subacute or chronic, as well as on inflammations of the larynx and pharynx, and it also excites in these parts a special action easily appreciable in a great number of cases, which changes the morbid condition, and causes the final disappearance of chronic affections established in one or several points of the respiratory system.

The patients, therefore, who will be sent to Eaux-Bonnes will be those attacked with laryngitis, and chronic, simple, or granular pharyngitis; bronchial catarrhs, both recent and of long standing; bronchial asthma, chronic pleurisy with effusion, &c.

In addition to the diseases above enumerated, there is still another for which the Eaux-Bonnes waters have been strongly recommended, namely, pulmonary phthisis. But here some explanations are indispensable. Without entering upon the theoretical discussions as to the nature, causes, and pathological development of tuberculosis, to which many medical men have lately devoted themselves, it will suffice to state the conclusions to which Dr Cazenave de la Roche has been led by seventeen years of practice at Eaux-Bonnes. "In certain morbid states of the respiratory organs, the nature of which is not clearly indicated by the symptoms, so that their precise diagnosis is difficult, and particularly in cases where the tubercle is still in the state of granulation, and is not detected by auscultation, the waters of Eaux-Bonnes may, by their searching action, render immense service as a sort of touch-stone. Secondly, taken in the first stage of torpid pulmonary phthisis (that is to say, in cases where the tubercle is in a state of crudity or miliary granulation, and the constitution is lymphatic or scrofulous), the waters of Eaux-Bonnes disperse the concomitant epigenetic congestion, and by their elective action on the respiratory organs occasion the disintegration of the anatomical element of the disease, and also combat the diathesis by their constitutional regenerative action. At other stages of torpid phthisis the waters of Eaux-Bonnes are impotent, when not positively dangerous; and they ought to be strictly forbidden whenever the tubercular affection assumes an acute form, whatever be the period of the disease."

It is an error widely diffused amongst practitioners that in most cases the waters of Eaux-Bonnes produce hæmoptysis. As a rule, the experience of all the medical men who practise at this Station is to the contrary, and they declare that if hæmoptysis is somewhat frequently observed among the patients at Eaux-

Bonnes, it is because it had already existed before
their arrival.

It is impossible to fix an exact duration for the
treatment, as it depends upon a number of circumstances which the local physician alone can rightly
appreciate.

Baths are little used at Eaux-Bonnes. Since the
construction of the Orteig and Hydrotherapeutic
Establishments, there are twenty private mineral-
water and ten fresh-water baths. This number, small
as it is, is amply sufficient for all requirements.

It is usual and advisable, after the season at Eaux-
Bonnes, for patients to complete their cure by going
to some seaside place, either to take sea baths, or
simply to enjoy the beneficial influence of sea air.

Of all sulphuretted waters, those of Eaux-Bonnes
(*Source Vieille*) are of the greatest value for exportation. It must, however, be borne in mind that as
they always lose a little of their strength in transport,
the dose generally taken is a full glass each morning.

Residence at Eaux-Bonnes is agreeable, and the
neighbourhood is very attractive. The *Horizontale*
promenade, which resembles a kind of balcony on the
side of the mountain, is a charming resort for those who
are not strong enough to climb the somewhat steep
slopes. It is not, however, planted with trees, and
therefore in the daytime visitors prefer the *Impératrice*
promenade, which runs along another slope of the
valley, and offers refreshing shade.

EAUX-CHAUDES

(*Basses-Pyrénées*)

SODIO-SULPHURETTED WARM OR COLD SPRINGS

THE approach to these springs was formerly as perilous as that to Eaux-Bonnes. It was necessary to climb a steep mountain, the *Hourat,* to pass through a narrow defile at the summit, cut out of the rock, and then redescend by a very steep slope. But now a fine wide road of really marvellous engineering work runs beside the torrent, and leads directly to the baths.

The village occupies the upper end of the Ossau Valley, which at this point forms a wild and sombre ravine. The houses are built close to the mountain, and on the borders of the torrent stands the Thermal Establishment, which is one of the finest in the Pyrenees.

The mineral waters, which are sodio-sulphuretted, are drawn from seven springs, of which the following are the names and temperatures :

Le Clot	.	.	.	.	36° C.	96·8° F.
L'Esquirette-chaude		.	.	34° C.	93·2° F.
L'Esquirette-tempérée		.	.		
Le Rey	.	.	.	.	33° C.	91·4° F.
Baudot	.	.	.	.	27° C.	80·6° F.
L'Aressecq		.	.	.	25° C.	77·0° F.
Minvielle		.	.	.	11° C.	51·8° F.

The springs of Le Clot, Esquirette-chaude, Esquirette-tempérée, and Le Rey, which furnish nearly 40,000 gallons per day, supply the Thermal Establishment.

The building, situated at the western foot of the Gourzy mountain, and on the right bank of the river Ossau, is composed of a square main building, entirely of white marble, with a courtyard in the centre. In this structure the visitors reside. Three semi-circular ranges of buildings are reserved for douches and bath-rooms.

The Baudot, L'Aressecq, and Minvielle springs supply three *buvêttes*, which are placed exactly at the point where the springs issue from the earth.

The waters of Eaux-Chaudes are clear, of a sulphurous odour, with a flavour of gall more or less pronounced, according to the spring, and they deposit in the reservoirs a quantity of barégine, which varies with the degree of sulphuration of each.

CHEMICAL ANALYSIS.

WATER 1000 grammes (1 litre).	SOURCE CLOT.
	Grammes.
Sulphur	0·003625
Hydrochloric acid	0·0561
Sulphuric acid	0·0811
Carbonic acid	0·0048
Silicic acid	0·0550
Potash	0·0079
Soda	0·0922
Lime	0·0284
Traces of iodic and boric acids, ammonia, lithia, magnesia, alumina, oxide of iron, and organic matter. —*Mialhe & Lefort.*	0·329125

The Eaux-Chaudes waters are stimulating in different degrees. Their action is chiefly exerted on the skin and mucous membranes. At the commencement of the treatment they usually cause an abundant diuresis, or perspiration, and sometimes even bring about *la poussée*, or thermal fever. They are also sedative, like those of Saint-Sauveur.

They are employed in baths, with great success, for certain kinds of rheumatism, rather muscular than articular, characterised by great irritability, and in which the nervous element plays the largest part; and thus used they are less likely to excite febrile complications. They have also been greatly extolled in cases of neuralgia. It is not clear why Bordeu calls them " strong and impetuous :" they only become so when taken in excess, which is also the case with the greater number of mineral springs. But what constitutes in a great measure the distinctive merit of the waters of Eaux-Chaudes, is their peculiar tendency to produce congestion of the uterus, and consequently to re-establish menstruation. Thus, it is very common in the case of chlorotic young girls to find the catamenia reappear after twelve or fifteen days' use of these waters, and in this respect they often act better and quicker than ferruginous springs.

Eaux-Chaudes offers the great advantage to patients that they can reside in the establishment where their treatment is carried on, and where everything is arranged in the most comfortable manner for their reception.

Among the objects of interest to be visited are the promenades of the *Château d'Argout*, the *Pont d'Enfer*, situated on the road from Eaux-Chaudes to Eaux-Bonnes, and cut out of the rock in the frightful ravine of Hourat; the celebrated grotto of *Eaux-Chaudes*, remarkable for its depth, its stalactites of curious and fantastic shapes, and for the presence of a torrent which issues from the rocky labyrinth in a noisy cascade; the lake of *Aule,* the picturesque valley of *Bious-Artigues,* the famous *Pic du Midi d'Ossau,* &c.

SAINT-CHRISTAU

(Basses-Pyrénées)

CALCAREO-SULPHURETTED COLD SPRINGS, CONTAINING COPPER.

SAINT-CHRISTAU is a little Thermal Station situated about 5 miles from Oloron, on the right bank of the river *Aspe*, at the foot of Mont Binet. Its mineral waters, which have only been used for about a dozen years, come from five springs, known as the *Source des Arceaux, Source du Chemin, Source de la Rotonde douce, Source de la Rotonde froide,* and *Source Sulfureuse.*

They supply two Establishments, where they are drunk, and also used in baths and douches. The analysis of the principal spring may be taken as a type.

CHEMICAL ANALYSIS.

WATER 1000 grammes (1 litre).					SOURCE DU CHEMIN.
					Cubic cent.
Oxygen	.	.	.	.	. 7·60
Nitrogen	.	.	.	.	. 24·80
					Grammes.
Carbonic acid, free	.	.	.	.	0·0036
Bicarbonate of lime	.	.	.	.	0·1600
,, magnesia	.	.	.	.	0·0641
Chloride of sodium	.	.	.	.	0·0301
,, calcium	.	.	.	.	0·0236
Sulphate of lime	.	.	.	.	0·0098
,, copper	.	.	.	.	0·00034
,, iron	.	.	.	.	0·0046
Phosphate of lime	.	.	.	.	0·0015
Silicate of lime	.	.	.	.	0·0140
Traces of bicarbonate of lithia, chloride of magnesium, iodide of sodium, carbonate of manganese, arseniate of lime, silicate of potash, borate of soda, and organic matter.—*Filhol.*					0·31164

This analysis shows that the waters of Saint-Christau contain the novel element of sulphate of copper, which is not to be met with in any other French mineral water. It has been affirmed that they owe their activity to the combination of copper and iron.

These springs are clear and almost colourless, except on rainy days, when it has been observed by Dr Tillot that the *Source des Arceaux* is a little cloudy, and has a sulphurous smell. They have a slight mouldy odour, and a faint styptic taste. In a physiological point of view, they are at first exciting, and afterwards sedative, solvent, tonic, and regenerative.

The waters of Saint-Christau are especially recommended in inveterate cases of scrofulous or syphilitic cutaneous affections, and also certain chronic ulcerations; they are eminently cicatrising. Pulverisation has been ingeniously introduced into the thermo-mineral treatment of this Station, and by means of this process glandular sore-throats, chronic laryngitis, and cases of chronic ophthalmia, such as blepharotitis, keratitis, &c., have been cured.

The environs of Saint-Christau are delightful, as, indeed, are those of all the Thermal Stations of the Pyrenees.

CAUTERETS

(*Hautes-Pyrénées*)

SODIO-SULPHURETTED WARM SPRINGS

CAUTERETS is a pretty little town of 1300 inhabitants, situated in the valley of *Lavedan*, beneath the shadow of a double chain of mountains. It was formerly a very difficult place to reach, but is now easily accessible by means of the fine road from Pau, which gives the traveller the opportunity of enjoying the picturesque scenery of the Argelès valley.

The waters of Cauterets were known to the Romans, and it is even alleged that the existing "Cæsar's Bath" was constructed by Julius Cæsar for his soldiers.

The springs are very numerous, and of temperatures varying between 25° and 60° C. (77° and 140° F.); they are divided into two groups:

EASTERN GROUP.			SOUTHERN GROUP.		
	C.	F.		C.	F.
César . . .	48°	118°	De la Raillière . . .	39°	102°
Espagnole . .	47°	117°	Petit St Sauveur (ancienne)	35°	95°
Pauze vieux . .	41°	106°	Du Pré . . .	49°	120°
Sulfureuse nouvelle	29°	84°	De Mahourat . . .	50°	122°
Du Rocher . .	40°	104°	Des Œufs . . .	59°	138°
			Des Yeux . . .	24°	75°
			Du Bois . . .	43°	109°

All these springs are scattered round the town, and for the most part are situated at considerable distances apart. They produce altogether more than 2,640,000 gallons of water a day, and supply several more or less important Establishments.

The principal of these, called *Thermes des Œufs,* is undoubtedly one of the finest in France. It contains all the resources and all the apparatus now considered indispensable for perfect hydromineral treatment, and is supplied by the *Source des Œufs,* which delivers 132,276 gallons of water per day. A splendid swimming bath, sixty-five feet long by twenty-six feet wide, constructed in a central garden, and supplied with running mineral water, is especially noteworthy.

The Grand Establishment, also called *Thermes de César* or *des Espagnols,* was built in the year 1844, on the designs of Artigala. It stands at the foot of the mountain Peyraute, at an altitude of 3000 feet above the level of the sea, and is entirely constructed of grey marble. It is supplied by two springs, and contains 130 polished marble baths, besides a large number of douches of various forms and temperatures.

There are also special places reserved for *pédiluves*, and rooms for inhalations and pulverisations.

The waters of Cauterets are clear, colourless, and do not "whiten" like those of Luchon; they have a smell and taste resembling those of gall, which vary in intensity with the different springs. According to M. Filhol, they give off less sulphuretted hydrogen than those of Luchon, and deposit no sulphur in the pipes. Several of them deposit barégine on coming into contact with the air, and all become easily decomposed, except the waters of *La Raillière* and *César;* which are exported without undergoing any alteration, thanks to a special method invented by M. Broca, by which the water is bottled without ever having come into contact with the air.

The Establishment of *La Raillière* takes its name from the spring that supplies it, and which is the most frequented of the Cauterets group. The *Thermes* form a parallelogram, built on a large terrace nearly 300 feet long. From this terrace one enters a gallery well lighted by lofty windows. In the middle of the edifice is the *buvette,* which is directly over the spring; and occupying the whole length are thirty bath rooms, fitted up with marble baths.

The legend of the place relates that the virtues of this water were pointed out to some shepherds by an emaciated cow, whom they observed drinking habitually at the stream, and gradually regaining plumpness and health.

CHEMICAL ANALYSIS

WATER 1000 grammes (1 litre).					SOURCE DE LA RAILLIÈRE.
					Litre.
Nitrogen	.	.	.	.	0·004
					Grammes.
Lime	.	.	.	.	0·004487
Magnesia	.	.	.	.	0·000445
Caustic soda	.	.	.	.	0·003396
Sulphide of sodium	.	.	.		0·019400
Sulphate of soda	.	.	.	.	0·044317
Chloride of sodium	.	.	.	.	0·049576
Silica	.	.	.	.	0·061097
Traces of barégine, caustic potash and ammonia.— *Longchamps.*					0·182718

La Raillière (like the *Source-Vieille of Eaux-Bonnes*) is prescribed in catarrhal and tubercular affections of the chest, but the action of these two waters differs in certain particulars which it is essential to explain.

The waters of *La Raillière* are not so exciting as those of *Eaux-Bonnes,* and occasion less risk of hæmoptysis. The cause of this must undoubtedly be sought in the difference of energy in their respective action; but perhaps the method of administration of the mineral water itself must also be taken into consideration.

It has been seen that at *Eaux-Bonnes* there is very little bathing; at *La Raillière,* on the contrary, the temperature of the spring and its abundant supply of water allow the daily use of complete or partial baths. In the latter, which are most frequently employed, the patient is seated in the bath, the chest and arms covered with flannel, and the water reaching to the

8

umbilicus. By thus attracting the blood to the skin and towards the sub-diaphragmatic region, a moderating and derivative influence is exercised on the congestive tendency, which the internal use of the mineral water might occasion in the respiratory organs, thereby sufficiently accounting for the greater rarity of hæmoptysis.

The water of *La Raillière* is to be preferred to that of the *Source de César* whenever a chronic affection of the mucous membrane of the air-passages coexists with chlorosis, anæmia, lymphatic affections, or scrofula. It has been employed with success in cutaneous affections of a scrofulous nature; against urethral discharges of long standing, and spermatorrhœa. It has also been useful in contending with the tendency to passive uterine hæmorrhage and profuse leucorrhœa.

The *Mahourat* spring is only employed for drinking purposes. It is situated in front of the beautiful cascade of the same name. Its temperature is 49° C. (120° F.), and its sulphurisation gram. 0·013. As the access to it is difficult and fatiguing to many invalids, a new *buvette* has been constructed at Benquès Bridge, so as to avoid the somewhat trying ascent, which patients were formerly obliged to make. This spring possesses an incontestable efficacy in the treatment of dyspepsia and gastralgia, especially when these affections are connected with an atonic condition, for it acts as a slight stimulant on the mucous membrane of the digestive organs. This explains why it is sometimes advantageously mixed with the water of *La Raillière*, when the latter is found to lie somewhat heavily on the stomach.

It has also been observed that the *Mahourat* water facilitates the elimination from the system of an excess of uric acid, and consequently is useful against gravel and renal calculi. This action, however, is common to many other sulphurous springs, and in this respect those of *La Preste* are preferable.

The springs *de César* and *des Espagnols* rise at an elevated point of the mountain called *Pic-du-Bain*, and from thence are conducted by an aqueduct of about 330 feet in length to the Grand Establishment already described.

The *Source de César* has a temperature of 46° C. (114·8° F.), and contains per litre 0·24 gram. of sulphuret of sodium.

The *Source des Espagnols* differs only from that of *César* in being less sulphurous and not so warm.

There are also the springs of *Le Pré*, *Le Bois*, and *Le Rocher*, which are of less importance.

The waters of Cauterets are drunk, and used for baths, douches, inhalations of vapour, and pulverisations. Among the diseases most successfully treated at this station the following may be mentioned: pharyngitis, chronic laryngitis, bronchitis, catarrh, asthma, emphysema, pleural effusion, phthisis, and in general all chronic diseases of the thoracic cavity; also dyspepsia, gastralgia, and chronic gastritis; uterine affections, scrofula, and syphilis.

It has been shown that the different springs of Cauterets combine nearly all the properties of the Pyrenean group of mineral waters, so that almost every kind of sulphurous medication finds an easy application here. It is not, however, to be supposed that they can replace all the others; for to any one who has studied hydrotherapeutics each thermal station, or even each spring, has a speciality of action on a certain class of diseases which can neither be attained or equalled by the waters of any other station, even though they appear to be chemically equivalent.

The waters of the springs *César*, *Mahourat*, and *La Raillière* are exported, and keep without undergoing any deterioration.

SAINT-SAUVEUR

(*Hautes-Pyrénées*)

SODIO-SULPHURETTED WARM SPRINGS

Two defiles begin at Pierrefitte; that on the right leads to Cauterets, and that on the left to Saint-Sauveur. It would be difficult to convey an idea of this latter route, boldly cut out of the rock, which it breaks through when it cannot skirt; supported by high arches that spring across the torrent, and passing seven times from one bank to the other over as many marble bridges, in order to find less refractory slopes.

Looking upwards, the traveller perceives far away between the two beetling sides of this frightful gorge, a narrow strip of sky. Below, the torrent is sometimes seen, and sometimes only heard, roaring at a depth which the eye cannot fathom. There is no habitation to be seen, no trace of cultivation; on all sides are bare, ragged, and rocky mountains, whose summits are as white and inaccessible as glaciers.

As the road approaches Luz, the double rampart formed by these mountains opens out, vegetation reappears, the fields are filled with life and movement; and soon the tourist finds himself among charming scenery. This is the valley of Luz. It takes only fifteen or twenty minutes to go on foot from the town of Luz to the village of Saint-Sauveur, crossing the torrent by a pretty marble bridge.

Saint-Sauveur is situated at a height of 2500 feet above the level of the sea, and is, as it were, suspended half way up the side of the *Laze* mountain. The present village consists only of a single street; but, small as it is, it was necessary to blast out of the solid rock a site on which to build it. The principal spring (called *les Bains*) rises in front of the Establishment, and into this it is conducted by underground pipes; the water is quite clear, and soft to the taste and touch. Its temperature at the rising of the source is 34° C. (93° F.), and it contains gram. 0·0217 of sulphide of sodium. The Thermal Establishment is a rectangular peristyle, ornamented with Corinthian columns, and affording a charming view of the Gavarnie torrent,

which it overlooks. Around the terrace there are
twenty well-fitted bath-rooms, two rooms for ascending and descending douches, and two *buvettes*.

CHEMICAL ANALYSIS

WATER 1000 grammes (1 litre).				SOURCE DES BAINS.
				Grammes.
Sulphide of sodium	.	.	.	.	0·0218
Chloride of sodium	.	.	.	.	0·0695
Sulphate of soda	.	.	.	.	0·0400
Silicate of soda .	.	.	.	.	0·0704
„ lime .	.	.	.	.	0·0062
„ magnesia	.	.	.	.	0·0031
„ alumina	.	.	.	.	0·0070
Organic matter .	.	.	.	.	0·0320
Traces of boric acid and iodine.—*Filhol.*				0·2500

The Saint-Sauveur waters give an unctuous sensation to the skin. They are beneficial in the treatment
of neuralgic affections, particularly facial neuralgia
and sciatica; and are marvellously efficacious in
those ill-defined nervous affections, called by the
French "Névroses," which are almost peculiar to
members of the upper classes of society, and are unknown to the working class, whose nerves are
strengthened or blunted by their arduous labours.

The waters of Saint-Sauveur are supremely efficacious in diseases of the womb, and that really constitutes their speciality. Under the influence of baths,
ascending douches, and vaginal injections, there is a
rapid disappearance of those congestions and granulations of the neck of the womb, which are so often

accompanied by leucorrhœa and relaxation of the
ligaments. The late Dr Fabas used to say that the
greater part of the patients left their pessary behind
them as an *ex-voto*.

Affections of the urinary organs are also benefited
by the waters of Saint-Sauveur, and in their case
drinking is associated with bathing. These waters
are especially successful in catarrhal affections of the
bladder, for which saline waters would be inefficacious
or even irritating; they render the urine softer and
more abundant, and modify the vitality of the mucous
membrane, restoring its secretion to a healthy condition. Finally, they may also be useful in reducing
chronic prostatic enlargements.

The waters of Saint-Sauveur, although highly
charged with *barégine*, are generally pretty well borne
by the stomach, a fact that must doubtless be attributed to the quantity of gas contained in them, and
which may be seen escaping and sparkling in the
glass.

The Thermal Establishment possesses only two
springs; but at a few minutes' walk from the village,
and upon a height which dominates it, there is another
sulphurous spring called *la Hontalade*, the temperature of which is 22° C. (72° F.), and its sulphurisation gram. 0·019 per litre. It is employed with
advantage against all forms of gastralgia, and is to
Saint-Sauveur what Mahourat is to Cauterets. Baths,
douches, and a *buvette* have also been established
there.

Saint-Sauveur is one of the most delightful thermal stations of the Pyrenees. The neighbourhood affords to bathers and tourists a great numbers of walks and picturesque excursions, among which may be specially mentioned those to the *Plateau de Bellevue*, the *Sassis* road, the little towns of *Luz*, *Barèges*, and *Cauterets;* the *Pic de Bergeons* and *Pic d'Aubiste;* the famous *Pic du Midi*, and the celebrated *Cirque de Gavarnie.*

BARÈGES

(*Hautes-Pyrénées*)

SODIO-SULPHURETTED WARM SPRINGS

BARÈGES, which is situated more than 4000 feet above the level of the sea, is neither a town nor a village ; it is a picturesque group of houses, cottages, and wooden huts, scattered about on the banks of the *Bastan*, an impetuous torrent which rushes from the summit of the mountain. Formerly this station was almost inaccessible, but now carriages drive to it along an excellent road.

The waters of Barèges were known to Julius Cæsar, who erected monuments there marked with the characteristic dignity of Roman works. Margaret, Queen of Navarre, and sister of Francis I, restored to these waters a part of the renown which they had enjoyed in antiquity. Henry IV knew them, and in his youth was a frequent visitor. Lastly, in 1675, Fagon, physician to Louis XIV, sent there, under the care of Madame de Maintenon, the Duke of Maine, the king's natural son, who suffered from a scrofulous affection. The relief which he experienced, and the letters which Madame de Maintenon wrote to Louis XIV, rendered Barèges more celebrated than ever, and the royal father endowed this watering place—almost lost in the midst of a wild, lonely country—with a military hospital. From that time this Thermal Station has enjoyed a great reputation.

CHEMICAL ANALYSIS

WATER 1000 grammes (1 litre).	
Sulphide of sodium	0·0159
Sulphate of soda	0·0202
Silicate of soda	0·0201
Chloride of sodium	0·0320
Iodide of sodium	0·0040
Sulphate and carbonate of iron . . .	0·0110
Chloride of magnesium	0·0400
Carbonate of lime ⎱ „ magnesia ⎰ . . .	0·0020
Silicates of alumina and lime . .	0·0110
Albuminous matter and glairine ⎱ Loss ⎰ . .	0·0120
Latour de Trie.	0·1653

The waters of Barèges are sodio-sulphuretted; they are limpid, clear, soft, and oily to the touch, and hardly exhale any sulphurous smell in the glass that holds them. They have a marked hepatic flavour, which leaves a dull and somewhat nauseous after-taste.

They contain a nitrogenous substance which Long-champ was the first to describe under the name of *Barégine*, and which is found in all the other sulphurous waters of the Pyrenees. They do not "whiten" on coming into contact with the air, as do those of Luchon; the reason, says M. Filhol, in his book on the Waters of the Pyrenees, is "that they do not contain an excess of silicic acid."

The Thermal Establishment has been constructed with due regard to the special climate of Barèges. All the bathing arrangements, baths, douches, hot-air baths, and *buvettes*, are grouped in one great hall, 170 feet long by 60 feet wide, with a high, open roof. It contains an immense waiting-room, and above the bathing apartments are galleries, in which are rooms for repose and conversation. Besides the civil and military *piscines*, which form a distinct group, the Establishment contains twenty-five baths, furnished with injection douches; also ascending douches, a warm *piscine*, or family bath, a complete service of vapour baths and douches, and four soft water or emollient baths.

The waters of Barèges are eminently stimulating, and when taken in doses of three or four tumblers-full, are easily absorbed. They quicken the circulation, and thus powerfully aid the effects of the bath, which forms the basis of the treatment. The waters of the springs of Tambour and Saint-Roch supply the *buvettes*.

The speciality of this Thermal Station consists in the

treatment of old traumatic affections; and few foreign substances, whether bullets or other projectiles, fragments or splinters of bone, resist the expulsive action of these waters. Under their influence, the unhealthy growth of flesh which so often occurs round the orifice of fistulous openings, becomes covered, after a few baths, with a whitish pellicle, extremely firm, and resembling in appearance that produced by superficial cauterisation with nitrate of silver. This pellicle subsequently falls off, and the tissues present a more healthy, life-like aspect. This phenomenon, which is reproduced after each bath, is attributed to the caustic action of the alkaline sulphur held in solution in the mineral water. While the exterior wound thus becomes modified, the internal surface of the fistulous cavity gradually contracts, and closes more and more on the foreign body or loose substance, which it gradually expels, when a healthy cicatrix covers the place which it occupied.

The waters of Barèges also give excellent results in the treatment of contusions, old sprains, profuse suppurations; caries of the bones, and necrosis; denudations of the tendons, muscular and tendinous contractions, imperfect cicatrisation, stiffness of joints, and enlargements consequent on fractures and dislocations. They are also employed with success against inveterate tertiary symptoms of constitutional syphilis, and in skin diseases specially connected with the herpetic diathesis.

These waters keep well, and are principally employed as lotions in certain diseases of the skin.

Residence at Barèges is only moderately lively, as the greater part of the bathers are scarcely in a condition for drawing-room amusements.

In the streets and on the promenades, one sees nothing but crutches, slings, military cloaks, and sedan chairs, scarcely promising auguries of cheerful society.

The environs of Barèges, however, are extremely picturesque; and its situation among wild and varied scenery offers to the numerous bathers, who each year visit this station, natural pleasures which are not met with elsewhere.

BARZUN-BARÈGES

(*Hautes-Pyrénées*)

The sodio-sulphuretted warm spring of Barzun rises near Bastan, at a little distance outside Barèges. Its temperature is 31° C. (87·8° F.), and its mineralisation gram. 0·038 of sulphuret of sodium.

The physical and medicinal characteristics of this water are the same as those of the Barèges springs.

A company has been formed to bring the water down to Luz, only a short distance from the source, and to erect there a suitable establishment, with baths, douches, and *buvettes*. The little town of Luz would profit by the enterprise, and some invalids would prefer to be saved the trouble of going up the mountain to Barèges.

BAGNÈRES-DE-BIGORRE

(*Hautes-Pyrénées*)

CALCAREO-SULPHATED, OR CALCAREO-SULPHURETTED, OR
FERRUGINOUS, WARM AND COLD SPRINGS.

BAGNÈRES-DE-BIGORRE is a charming town of 10,000 inhabitants, situated on the banks of the *Adour*, at the extremity of the beautiful plain of Tarbes, at an altitude of 1800 feet above the level of the sea.

The mineral waters of Bigorre may be divided into three classes : sulphated, sulphuretted, and ferruginous springs.

The Thermal Establishments are fourteen in number. Thirteen belong to private persons, while the fourteenth (which is the largest and most important) is the property of the town, and bears the name of Marie-Thérèse. It is built of marble, and close behind it rises the mountain from whence come the springs which supply it. It includes thirty-six bath-rooms with dressing-rooms attached, furnished with spacious baths and small douches for local applications ; also douche-rooms, vapour baths, public baths, and *buvettes*.

These fourteen Establishments are supplied by a large number of springs, which are all calcareo-sulphated.

The waters of Bagnères-de-Bigorre are laxative and diuretic, especially those of the *Reine*, the *Serre*, and the *Salut* springs. The hottest springs are stimulating and exciting, but the tepid ones are sedative, and slightly weakening. Iron is found in considerable quantity in the former, while, according to the analyses of M. Filhol, the latter would seem not to contain any. As it would be impossible, for want of space, to give the analyses of all the springs, that of *La Reine* may be taken as a type, being the richest and most important of the group.

CHEMICAL ANALYSIS

WATER 1000 grammes (1 litre).				SOURCE LA REINE.
Chloride of magnesium	.	.	.	0·130
,, sodium	.	.	.	0·062
Sulphate of lime	.	.	.	1·680
,, soda }	.	.	.	0·396
,, magnesia }				
Carbonate of lime	.	.	.	0·266
,, magnesia	.	.	.	0·044
,, iron	.	.	.	0·080
Resinous fatty matter	.	.	.	0·006
Silicic acid	.	.	.	0·036
Loss	.	.	.	0 054
				2·754

The sulphurous springs of Bagnères-de-Bigorre are two in number, the *Pinac* (calcareo-sulphuretted), and *Labassère* (sodio-sulphuretted). The first, situated in the town itself, has a temperature of 18·7° C. (65·75° F.) ; the second, nearly five miles distant from Bagnères, only attains a temperature of 13·8° C. (56·75° F.). The Pinac spring supplies six baths and two *buvettes*. The water of Labassère is only used for drinking, and can be exported.

The ferruginous springs are three in number : the *Angoulème*, the *Brauhaubant*, and the *Rousse*.

The Bagnères waters are successfully used in the treatment of numerous diseases, especially certain uterine affections characterised by hyper-sensibility. They are also employed in cases of anæmia, chlorosis, and leucorrhœa; in pathological sequelæ of intermittent fever, in certain forms of paralysis, in caries of bones, and necrosis. Excellent results are obtained in chronic rheumatism, neuralgia, and chorea, nervous palpitations and gastralgia; also in certain skin diseases for which strong sulphurous waters would be too active; and, finally, they have proved beneficial in catarrhal or tubercular affections of the lungs.

Bagnères-de-Bigorre and the surrounding country offer many attractions to visitors. The Marie-Thérèse Establishment provides billiard, conversation, and reading-rooms, &c., while exercise may be taken by visiting the beautiful promenades of the *Coustous*, the *Vigneaux*, the *Allées-Maintenon*, the *Fontaine-Ferrugineuse*, the *Salut* baths, &c. Those who like to make excursions on horseback, or in carriages, will not fail to go to *Lourdes*, the *Escaledière*, the marble quarry of *Campan*, the Valley of *Lesponne*, or to make the ascent of the *Pic-du-Midi*, *Mont-Aigu*, and the *Lac-Bleu*.

CAPVERN

(*Hautes-Pyrénées*)

CALCAREO-SULPHATED WARM OR COLD SPRINGS

CAPVERN is a little village of 900 inhabitants, built on a hill at a height of 1300 feet above the level of the sea, about five miles from Lannemezan. The springs, which are two in number, rise at some distance from the village in two deep glens. The first bears the name of the *Hount-Caoude* spring, the second is called the *Bouridé*. A few years ago the waters of Capvern were little frequented, and only by a few patients from the neighbourhood, but the number of visitors has gradually increased, and at present it exceeds 3000 per annum.

CHEMICAL ANALYSIS

WATER 1000 grammes (1 litre).	
	Cubic cent.
Carbonic acid	49
Oxygen	18
Nitrogen	28
	—
	95
	Grammes.
Sulphate of lime . . .	1·096
„ magnesia . . .	0·464
„ soda . . .	0·072
Chloride of magnesium . . .	0·032
„ sodium . . .	0·044
„ calcium . . .	0·016
Carbonate of magnesia . . .	0·012
„ lime . . .	0·220
„ iron . . .	0·024
Silica	0·028
	2·084

The waters of Capvern are calcareo-sulphated. The two springs supply in twenty-four hours nearly 66,140 gallons of water, at a temperature of about $24\cdot37°$ C. (77° F.) The waters are perfectly clear, harsh to the touch, without smell, and with a sweet flavour, but which leaves behind a somewhat salt after-taste. They deposit a slight sediment of an ochreous colour, as noted by M. Latour, and allow the spontaneous and continuous escape of a colourless gas.

The waters are employed in all ways, but especially for drinking. They agree particularly well with the stomach, as is shown by the surprising appetite that is developed under their influence. The action of this water is very soon brought to bear upon all the digestive organs, the functions of which it powerfully stimulates. At the same time all the signs of congestion of the whole vascular system below the diaphragm manifest themselves, and consequently a renewal of activity ensues in all the corresponding organs. The urine is accordingly augmented in quantity, the bile becomes more abundant and more fluid, and the mucous membrane of the bowels increases its secretions. But what may be called the "crisis" is chiefly localised about the plexus of hæmorrhoidal veins, which are then swollen and congested, unusual throbbings are felt, and they become the centre of a hæmorrhagic flow.

This being the mode of action of the Capvern waters, it will be easily understood that they are suit-

9

able for a large number of diseases of the urinary
passages (gravel and vesical catarrh), certain disorders
of the liver and bile-ducts (hypertrophy and hepatic
calculi), many affections of the uterus (dysmenorrhœa,
amenorrhœa, leucorrhœa) ; and finally, that they prove
most efficacious remedies in that particular state
which is no longer health, without being exactly disease, and which arises from hæmorrhoidal tendency;
for it is most remarkable that the Capvern waters
often increase—but sometimes, on the contrary, sup-
press—the hæmorrhoidal flux. On this point the
following explanation has been given by M. Corties :—
"The waters act here as alteratives, by favouring the
development of the hæmorrhoidal symptoms, give a
free course to the blood, and relieve the overcharged
venous system of the abdomen and the surrounding
tissues. The circulation, which was slow and embar-
rassed, is quickened, and the disease disappears with
the cause which produced it."

One consequence, not less important, of these
effects upon the abdominal circulation, is the relief of
the upper parts of the body by the diversion of the
blood which would have a tendency to fix itself there.
Thus certain obstinate headaches, an habitual congestive state of the brain, buzzing in the ears ; cases of
hæmoptysis, and certain exanthemata of the face, have
more than once yielded to the derivative action of
these waters.

Life at Capvern is easy and not expensive, and
there is plenty of recreation for those who like walks

and picturesque excursions, as the Pyrenees are close at hand.

BAGNÈRES-DE-LUCHON

(*Haute-Garonne*)

SODIO-SULPHURETTED WARM SPRINGS, FERRUGINOUS COLD SPRINGS.

BAGNÈRES-DE-LUCHON is a little town of 4000 inhabitants, built on the banks of the river Pique, at the most westerly corner of the valley which bears its name, and at a height of about 2000 feet above the level of the sea. The valley in which it stands is one of the most magnificent in the Pyrenees.

Of its numerous mineral waters, some are sodio-sulphuretted and others ferruginous. The sodio-sulphuretted springs are fifty-three in number, divided into lower springs (*sources inférieures*), which rise in the interior of the Establishment itself, and upper springs (*sources supérieures*), which are situated about half a mile from the Establishment, to which they are conducted through subterranean galleries and earthenware conduits.

It is unnecessary to give here the detailed chemical analysis of these numerous springs; the two following may be taken as typical:

CHEMICAL ANALYSIS

WATER 1000 grammes (1 litre).			SOURCE LA REINE.	SOURCE DE LA GROTTE SUPÉRIEURE.
			Grammes.	Grammes.
Sulphide of sodium	.	.	0·0550	0·0314
,, iron	.	.	0·0028	0·0027
,, manganese	.	.	0·3300	0·0013
Sulphate of potash	.	.	0·0087	0·0059
,, soda	.	.	0·0222	0·0682
,. lime	.	.	0·0323	0·0682
Chloride of sodium	.	.	0·0674	0·0723
Silicic acid .	.	.	Traces	0·0103
Silicate of soda	.	.	Traces	0·0094
,, lime	.	.	0·0118	0·0376
,, magnesia	.	.	0·0083	0·0057
,, alumina	.	.	0·0274	0·0109
Alumina .	.	.	Traces	0·0109
Traces of carbonate of soda, sulphide of copper, hyposulphite of soda, iodide of sodium, phosphates, sulphuretted hydrogen, and organic matter.			0·2671	0·2557

The Thermal Establishment is one of the finest in existence. It is built on the site of the old Roman *thermæ*, at the foot of the mountain of *Super-Bagnères*, and consists of eight pavilions, containing the bath-rooms, douches, public and vapour baths, inhalation and pulverisation rooms.

The general arrangements are so good that the invalid, from his entrance into the Establishment until he reaches his bath, passes through an atmosphere increasingly warm and charged with sulphurous vapours, and on leaving the bath he finds these conditions reversed, so that the heat and sulphurous evaporation gradually diminish. Thus he reaches the outer air and regains his dwelling without any abrupt or dangerous transition. After each round of baths, (*i.e.* every hour and a quarter), there are available 120 baths, each provided with a douche for local application; from twenty to thirty places in the smaller public baths, which are four in number, and thirty admissions into the large swimming-bath. There are also five large douches, thirtý descending ones, falling into the same number of baths; hot-air and vapour-baths for forty patients; special rooms for inhaling both vapour and pulverised water, &c.

At the back entrance of the Establishment are nineteen reservoirs, into which the waters are brought by conduits made of glazed earthenware, and whence they are distributed to the bath-rooms by tubes of kyanized wood, which have been adopted as better suited to preserve their temperature.

An inscription placed in each pavilion indicates the group of springs which supply it. The following table gives the name, temperature, and sulphurisation of each of these groups :

SOURCES.	Temperature.	Sulphuret of sodium.
		Grammes.
1st Group { Bosquet .	. 38° C. 100° F.	0·031
Bordeu .	. 42° C. 108° F.	0·039
2nd Group { Étigny .	. 41° C. 106° F.	0·033
Ferras . .	. 38° C. 100° F.	0·018
3rd Group { Blanche . .	. 37° C. 99° F.	0·016
Reine .	. 44° C. 111° F.	0·031
Grotte Intérieure .	. 55° C. 131° F.	0·063
4th Group { Richard Supérieure	. 48° C. 118° F.	0·047
Richard Inférieure .	. 38° C. 100° F.	0·021

Each group, being thus composed of springs varying

in strength, is distributed among the baths by separate taps; and thus, according to the proportions in which the waters are mingled, the baths can be made weaker or stronger.

The large douches, each in a separate room, are supplied by the united reservoirs of *Bayen, Grotte Supérieure, Azémar,* and *Borden.*

Lastly, there is the *humage* or direct inhalation of the sulphurous vapours, which does not appear to be employed anywhere but at Luchon.

Such are the excellent arrangements provided for external treatment. The *buvettes* are also distributed into four groups as follows :

CHEMICAL ANALYSIS.

BUVETTES IN THE ESTABLISHMENT.		TEMPERATURE	SULPHIDE OF SODIUM.
		Centigrade.	Grammes.
1st Group	Les Romains . . .	47°	0·051
	Ferras inférieure, No. 1 . .	34°	0·052
	Ferras inférieure, No. 2 . .	39°	0·039
BUVETTES OUTSIDE THE ESTABLISHMENT.			
2nd Group	Grotte supérieure . .	53°	0·044
	Reine . . .	49°	0·054
	Blanche . . .	39°	0·022
3rd Group	Ferras ancienne . .	28°	0·004
	Enceinte . . .	42°	0·058
	Ferras nouvelle . .	31°	0·011
4th Group	Pré, No. 3 . . .	40°	0·031
	Pré, No. 2 . . .	44°	0·058
	Pré, No. 1 . . .	51°	0·078
	Pré, No. 1, cooled . .	22°	0·071

The physiological effects of the waters of Bagnères-de-Luchon differ according to the springs. Some,

such as the *Grottes Supérieure* and *Inférieure,* are stimulating; others, on the contrary, such as *Ferras, Bosquet,*
and *Blanche,* are calming and sedative. These waters
are employed successfully in a large number of diseases classified under one or the other of the four
following broad divisions :

1st. *Skin affections.*—Chronic eczema, either local
or general, impetigo, lichen, psoriasis, ichthyosis,
acne, &c.

2nd. *Scrofulous affections.*—Glandular enlargements,
strumous adenitis, ulcers, fistulous openings, caries,
necrosis, arthritis, white swellings (chronic affections
of the joints), coxalgia, &c.

3rd. *Rheumatic affections.*—Chronic rheumatism,
rheumatoid arthritis, spurious anchylosis, rheumatic
paralysis, muscular contraction, &c.

4th. *Syphilitic affections.*—Constitutional syphilis,
secondary and tertiary symptoms, syphylodermata, and
the constitutional effects of the abuse of mercury, &c.

These waters are also employed with advantage in
catarrhal affections of the respiratory organs, granular
pharyngitis, chronic bronchitis, chronic pulmonary
catarrh ; in leucorrhœa, passive congestion of the neck
of the womb; spermatorrhœa, impotence consequent
upon onanism or venereal excesses, &c.

The list of complaints for which ferruginous waters
are beneficial is too well known to need repetition.

Bagnères-de-Luchon is one of the most agreeable
of all the health resorts in the Pyrenees, if, indeed, it
may not be given an unqualified preference. The

casino, balls, concerts, a museum of Pyrenean objects of interest, and a public library, constitute the attractions of the town. For visitors who like short and easily attainable walks, may be mentioned the avenues of *Étigny, la Pique, les Platanes,* and *les Soupirs,* the woods of *Super-Bagnères,* the *Fontaine d'Amour,* the garden of the Curé of *Montauban,* a little village which adjoins Luchon, &c. Those, however, of a more adventurous disposition, who are lovers of beautiful scenery, ought to make excursions to the *Vallée du Lys,* the *Lac d'Ao, Castelvieil,* the *Port de Vénasque, Bossost, Saint-Bertrand-de-Comminges,* the marble quarries of *Saint-Béat,* and *la Maladetta,* the *Grottes du Chat* and *Gargas,* &c.

SIRADAN

(*Hautes-Pyrénées*)

CALCAREO-SULPHATED SPRINGS, FERRUGINOUS SPRINGS.

Siradan is a village situated about twelve miles from Bagnères-de-Luchon, at the entrance of a bare and arid valley, notable only for its mineral springs. These are four in number; two are calcareo-sulphated, warm, and two cold ferruginous. The former are taken as drinks, and used for baths and douches in an elegant Thermal Establishment, which is comfortable and well managed; the second are used only for drinking.

CHEMICAL ANALYSIS

WATER 1000 grammes
(1 litre).

	Cubic cent.
Carbonic acid, free . . .	18
	Grammes.
Bicarbonate of lime	0·2000
,, magnesia	0·0255
Sulphate of lime	1·3600
,, magnesia	0·2800
,, soda	0·1090
Chloride of calcium	0·0500
Traces of chloride of potassium, sodium, calcium, magnesium, oxide of iron, silica, iodine, phosphate of lime, and organic matter.—*Filhol.*	2·0255

The calcareo-sulphated waters of Siradan are laxative, diuretic, and tonic. They are successfully employed in dyspepsia, gastralgia, abdominal plethora, gravel, vesical catarrh, and generally in all chronic affections of the urinary passages.

The ferruginous waters are suitable to the different diseases for which iron is usually recommended.

Siradan is very much frequented by the inhabitants of the Pyrenees.

ENCAUSSE

(*Haute-Garonne*)

CALCAREO-SULPHATED TEPID SPRINGS

ENCAUSSE is a village of 700 inhabitants, situated in the district of Saint-Gaudens, at the entrance of the

Valley of Cabanac, beneath the mountains of Sauve-
terre and Kagire.

The waters of Encausse are tepid calcareo-sulphated.
They rise at three different points, and are there known
respectively as the *Grande-Source, Petite-Source* (which
are communal property), and *Source-Dargut,* named
after the proprietor.

The temperature of these springs is 22° C. (71·6° F.)
They yield daily about 18,000 gallons of a clear,
colourless water, with no smell, and of a slightly
bitter taste. Numerous bubbles of gas rise through
the water, and burst at the surface.

CHEMICAL ANALYSIS

WATER 1000 grammes (1 litre).	SOURCE GRANDE ET PETITE.
Nitrogen and oxygen ⎰	Litre
Carbonic acid ⎱	0·005
	Grammes.
Carbonate of lime	0·0270
,, magnesia	0·0155
Sulphate of soda	0·0204
,, lime	2·1390
,, magnesia	0·5420
Chloride of sodium	0·3202
Silicic acid	0·0100
Iodine	0·0100
Traces of sulphate of potash, silicate of soda, oxide of iron, arsenic and organic matter.—*Filhol.*	3·0741

The waters of Encausse can be drunk, and also used
in baths and douches. They act specially on the
gastro-intestinal and genito-urinary organs ; on the
mucous membranes, the liver, and the vascular system
generally. Like certain German waters, they are at

once laxative, diuretic, and sudorific ; their action is equally tonic, restorative, and alterative.

Among the complaints against which the Encausse waters seem to act most efficaciously, special mention may be made of intermittent fevers ; uterine affections accompanied by nervous irritability and inflammatory symptoms ; hysteria and chlorosis. Congestions of the liver, spleen, and abdominal viscera, are also favourably affected by these waters.

AULUS

(*Ariège*)

CALCAREO-SULPHATED TEPID SPRINGS

AULUS is a little town of 950 inhabitants, enclosed by lofty mountains, in a valley watered by the river *Garbet*. It possesses two small Thermal Establishments, supplied by four springs, which yield about 19,900 gallons in twenty-four hours. The waters have a temperature of 20° C. (68° F.) ; they are clear, colourless, rather bitter, soft and oily to the touch.

The waters of Aulus are laxative, tonic, and diuretic. They stimulate the functions of the skin, and often induce hæmorrhoidal congestion. They are efficaciously used against debility of the stomach or intestines, vesical catarrhs, chlorosis, and anæmia. They have been said to cure inveterate syphilitic affections.

CHEMICAL ANALYSIS

WATER 1000 grammes
(1 litre).

				Cubic cent.
Carbonic acid, free				125
				Grammes.
Sulphate of lime	.	.	.	. 1·980
„ soda	.	.	.	. 0·100
„ magnesia	.	.	.	. 0 300
Bicarbonate of lime	.	.	.	. 0·097
„ magnesia .	.	.	.	. 0·043
Chloride of sodium ⎤				
„ calcium ⎬.		.		0·040
„ magnesium ⎦				
Silica, alumina and phosphates ⎱				
Iodine ⎰ .				0·080
Oxides of iron and manganese	.	.	.	0·005
Traces of carbonate of potash, organic matter and arsenic in the ochreous deposit.—*O. Henry.*				2·645

The environs of Aulus are peculiar and interesting. Bathers can make excursions to the lake *of Hers*, the towers of *Castelminier*, the mines of *Core* and *Argentières*, the lakes of *Garbet* and *Aube*, the iron mines of *Rancié*, &c.

The waters of Aulus are principally exported, and keep without any deterioration.

AUDINAC

(*Ariège*)

CALCAREO-SULPHATED FERRUGINOUS TEPID SPRINGS

THE baths of Audinac are situated about three miles from Saint-Girons, on a small hill which the visitor reaches through a beautiful avenue of plane trees.

Although these baths have long been known to the inhabitants of the country, it is only within the last few years that they have been frequented by strangers.

The waters are calcareo-sulphated and ferruginous, and have a temperature of 22° C. (71·6° F.) They flow from two springs, which yield together nearly 65,000 gallons in the twenty-four hours.

The Thermal Establishment, lately rebuilt, includes thirty bath-rooms, and six douches of various descriptions.

CHEMICAL ANALYSIS

WATER 1000 grammes (1 litre).	SOURCE DES BAINS.
	Grammes.
Chloride of magnesium .	0·008
Carbonate of lime .	0·200
,, magnesia .	0·010
Sulphate of lime .	1·117
,, magnesia .	0·496
Oxide of iron .	0·003
,, manganese .	0·008
Silicate of soda .	0·020
Organic matter .	0·042
Carbonic acid .	0·079
Traces of sulphide of calcium, iodide of magnesium, creuate of iron, alumina and silicate of potash.— *Filhol.*	1·983

The waters of Audinac are laxative, purgative, and diuretic. They are principally recommended for functional affections of the alimentary canal, vesical catarrhs, and calculi of the liver and kidneys.

Field sports are the only amusement which visitors can procure at this Station.

USSAT

(*Ariège*)

CALCAREO-SULPHATED AND BICARBONATED
WARM SPRINGS

Ussat is a village situated between Foix and Tarascon, in a pleasant valley, above which tower granite mountains completely bare, and through which flows the river Ariége. The Thermal Establishment, which stands in the middle of a handsome park, was completely reconstructed about twenty years ago, according to the plans of the able hydrographic engineer, M. François. It is in the form of a long gallery of the Doric order, and comprises forty-four rooms fitted with Carrara marble baths, several large public baths, halls for "grand" douches, two rooms for ascending vagino-uterine and rectal douches, three rooms furnished with hip-baths, also a vapour-bath, and inhaling rooms. As many as forty thousand baths could be given at Ussat in one season, and during the last few years a total of nearly 30,000 have been given to about 1500 patients.

The bathing system used is extremely convenient and ingenious. Each bath is furnished with a tap communicating with the reservoir called the *bath gallery*, and can be opened from it; the supply is, therefore, immediate and continuous. In this simple and ingenious manner, the water can be distributed to all the baths at the same moment if it is desired, as is done regularly every hour during the season, or else separately, according to the requirements of the patients.

The directors of the establishment have built a small hospital, which receives regularly every year 200 poor patients, who are boarded and lodged free of expense, and attended gratuitously by the official Inspecting Physician.

The warm mineral springs of Ussat have their source in the heart of a mountain, which is calcareo-jurassic at the base. They consist of a number of streamlets of mineral water, emerging from one common reservoir or subterranean lake, and traversing the layers of a black, stratified schist. These streams are secured by means of several catches, and conducted by four adits into the common reservoir called the *bath gallery,* to which they furnish daily about 275,000 gallons of a clear, colourless, and inodorous water, slightly astringent and bitter in taste, and oily to the touch. Their normal temperature is 38° C. (100° F.)

It follows from the chemical analysis that the waters of Ussat are calcareo-bicarbonated and calcareo-sulphated, faintly mineralised. They are soothing and sedative to the nervous system, when employed at a moderate temperature; but, on the contrary, when used too hot, they become markedly exciting, especially to patients of high nervous susceptibility.

CHEMICAL ANALYSIS

WATER 1000 grammes (1 litre).	
	Cubic cent.
Carbonic acid . .	16·57
Nitrogen . . .	20·38
Oxygen . .	1·05
	38·00
	Grammes.
Carbonate of lime . . .	0·6995
„ soda . .	0·0381
Sulphate of magnesia . . .	0·1791
„ soda . . .	0·0583
„ potash . . .	0·0200
„ lime . . .	0·1920
Chloride of magnesium . . .	0·0420
Organic matter and loss . . .	0·0471
Traces of carbonate of magnesia and iron.—*Filhol*.	1·2761

All affections of the womb come under the peculiar
speciality of these waters ; and the medical testimony
concurs as to the excellent effects derived from their
use in ulcerations of the interior of the neck of the
uterus and its external surface ; in amenorrhœa and
dysmenorrhœa ; hypertrophy of the body of the womb,
and particularly of the neck, and the different varieties
of uterine displacements.   Their sedative action is extremely useful in the treatment of chronic metritis,
which often becomes a proximate cause of nearly all nervous affections in females, in consequence of the reflex
action of this abnormal subinflammatory state upon
the sympathetic nerves ; and finally, they are indicated
in spasms of the vagina and neck of the womb ;
hysteria, nymphomania, and every kind of hysterical

neuralgia. The diseases of women do not constitute the sole class of cases treated at Ussat. Persons of the other sex suffering from chronic inflammation of the digestive and urinary organs also take these waters with advantage. M. Bonnaus has for some years treated chorea there with great success, and especially in cases in which it had already resisted all ordinary treatment.

The environs of Ussat are very interesting. Visitors ought to see the *Grotte de Sabat*, for a long time the haunt of coiners ; the *Grotte de Lombrive* or *des Échelles*, celebrated for the feats of a band of brigands who infested this neighbourhood at the beginning of this century ; the village of *Ornolac*, the church of which possesses a bell dating from 1093, and the ruins of the old *Château-fort de Tarascon*, &c.

AX

(*Ariège*)

SODIO-SULPHURETTED WARM OR COLD SPRINGS

Ax is a little town containing 2000 inhabitants, situated at the southern extremity of the Ariège Valley, about twenty-six miles from Foix, and at the confluence of three streams, the *Ascou*, *Orlu*, and *Mérens*, at a height of about 2400 feet above the level of the sea.

The thermal springs of Ax belong to the sodio-sulphuretted class, and are more than sixty in number.

They vary so greatly in temperature and composition
as to form a scale passing almost insensibly through
every degree of heat and sulphurisation. Only forty
of them are utilised, and these are divided between
three different Establishments :—*Le Teich, Le Couloubret,* and *Le Breilh.* We shall take the first as a type
of the Ax waters, which yield altogether about 103,000
gallons daily.

FIRST GROUP DU TEICH.

NAMES OF THE SOURCES.	TEMPERA-TURE.
	Centigrade.
Viguerie	73°
Grande Pyramide . . .	65°
Astrié chaude	52°
Astrié froide	22°
Quod	64°
De la Grotte	50°
Isabelle	55°
De l'Eau Bleue	48°
No. 6	38°
Patissier	36°
No. 4	46°
De la Pompe	19°
St. Roch à droite . . .	42°
„ à gauche . . .	36°
Joly (newly discovered) . . .	73°

The lately constructed Establishment of Le Teich is
the most important in this Station, on account of its
comfortable arrangements and the large number of
springs which supply it. It comprises four principal
buvettes, supplied by four springs, of successively
diminishing degrees of sulphurisation. Six springs
supply the bathing department, which consists of fifty-
two black marble baths, fitted up in clean and well

ventilated rooms. There are also four large douches, six ordinary fixed and several moveable ones, four pharyngian douches, and several pulverisers; lastly, two vapour-baths, and an inhalation-room.

The Establishment of Le Couloubret is the oldest and now the least frequented of the three; it is supplied by eleven springs, which are very varied in temperature and sulphurisation, yielding altogether 72,625 gallons daily. It contains two *buvettes*, and twenty-eight baths.

The smart and well-managed Establishment of Le Breilh is supplied by thirty-five springs, furnishing 30,500 gallons every day. It possesses several *buvettes*, twenty bath-rooms, containing nine baths of black marble, seven of copper, and eight of granite; two rooms for douches, and a hot-air room.

CHEMICAL ANALYSIS

WATER 1000 grammes (1 litre).	SOURCE VIGUERIE.
	Grammes.
Sulphide of sodium	0·0200
Chloride of sodium	0·0350
Sulphate of soda	0·0318
Silicate of soda	0·1102
„ lime	0·0185
„ magnesia	0·0006
Organic matter	0·0450
Oxide of iron	0·0002
Alumina	0·0001
Traces of phosphoric and boric acids, iodine, sulphide of potassium, and lithium.—*Garigou*.	0·2614

The Ax waters are used in several ways; for purposes of drinking, and for baths, douches, and vapour-baths.

Their physical and chemical properties are much the same as those of the waters of Bagnères-de-Luchon. With regard to their physiological and therapeutical effects, they have been divided by MM. Astrié and Garrigou into strong, medium, and mild springs. This classification, however, has not met with approval, as it is deficient from a clinical point of view; and that formulated by Dr Alibert is decidedly to be preferred. He has divided the Ax waters into two principal groups, the sodio-sulphuretted and the degenerated sulphureous springs, according as the sulphur is found in a state of monosulphide of sodium, or has already undergone transformations.

The sodio-sulphuretted springs are highly stimulating. They excite the whole system, inducing a remarkable activity of all the functions, producing a more or less intense febrile condition, agitation, insomnia, painful dreams, and a marked excitation of the digestive and genito-urinary organs.

The degenerated sulphureous springs are sedative and calming, often soothing the *thermal* excitement produced by the stronger treatment. They are also tonic and restorative.

The Ax waters are suitable for treating rheumatism, scrofula, certain skin diseases, and some nervous and subinflammatory affections, belonging for the most part to one of these diatheses.

Among other morbid conditions that can be relieved or cured by the sulphurous treatment at Ax may be mentioned catarrhal affections (bronchitis, chronic coryza, dry and humid asthma, &c.), traumatic and surgical diseases, lesions following on old wounds, fractures, sprains, and dislocations; remote symptoms of syphilis, chlorosis, anæmia, chloro-anæmia, &c.

The neighbourhood of Ax offers some excursions of interest, among which may be noted those to the *Monts de la Serre* and *Bonascre*, and the cascade of *Orlu.*

OLETTE

(*Pyrénées Orientales*)

SODIO-SULPHURETTED WARM SPRINGS

OLETTE is a village built close under the mountain, upon the left bank of the river *Tet*, about 1900 feet above the level of the sea. Its mineral waters, which are sodio-sulphuretted, were discovered towards the close of the last century. They proceed from at least thirty springs, which are divided into three groups, and known by the names of : *Groupe de la Cascade, Groupe de Saint-André,* and *Groupe de l'Escalade.* Their temperature varies from 27° to 78° C. (80·6° to 172° F.), but their chemical constitution is almost identical. The Olette waters are drunk, as well as used in baths and douches, and inhaled both as vapour and spray, in a very comfortable Thermal Establishment, built some years ago. They can be prescribed with advantage in rheumatism, general nervous affections, and sciatica ; chronic bronchial and laryngeal affections ; indolent ulcers, skin diseases belonging to the herpetic diathesis, and also swellings of a scrofulous nature.

This Thermal Station is not yet as well known as it deserves to be; but visitors will find it an agreeable place of residence, in the midst of a mountainous country abounding in picturesque walks and excursions. It has at present no great interest except for residents in the locality.

ESCALDAS

(*Pyrénées Orientales*)

SODIO-SULPHURETTED WARM SPRINGS

ESCALDAS is a hamlet belonging to the commune of Villeneuve, close to the Spanish frontier, and situated on a plateau overlooking the whole valley of the *Cerdagne*. Its mineral waters, which are sodio-sulphuretted, flow from three springs. The principal of these, called *Grande-Source*, yields nearly 176,000 gallons daily of a clear water, oily to the touch, smelling of sulphur, and with a slight flavour of gall. Its temperature is not less than 46° C. (115° F.) The two other springs are the *Source-Merlat* and the *Source-Tartère*.

The persons who frequent Escaldas as patients, and the language spoken there, render it rather a Spanish than a French watering-place.

SAINT-THOMAS (*Pyrénées Orientales*).—These are

warm sodio-sulphuretted springs, which might be used to considerable advantage, if they were situated in a country less rich in sulphurous waters.

MOLITG

(*Pyrénées Orientales*)

SODIO-SULPHURETTED WARM SPRINGS

MOLITG is a village of 600 inhabitants, situated in the Pyrenees, about six miles from Le Vernet, on the left bank of the river *Tet*. Its mineral waters issue from ten springs, which rise at a short distance from the village, on the brink of a ravine called *Torrent de Riell*, nearly side by side, and supplying two Establishments, very well kept, but not sufficiently extensive for the number of bathers. The temperatures of the first four vary from 35·2° to 37·75 C. (95·5° to 100° F.) These flow to the *Lhupia* Establishment; while the six others, the temperatures of which vary from 21° C. (70° F.) to 36·15° C. (98° F.) supply the óther Establishment, called the *Bains Mamet*.

The springs of Molitg belong to the sodio-sulphuretted class. The waters are clear when they first issue from the ground, but become slightly turbid on contact with the air, and give off bubbles of gas, oxygen, nitrogen, and carbonic acid. They have a slightly sulphurous smell and an alkaline taste, are oily to the touch, and deposit glairine in the pipes and reservoirs.

CHEMICAL ANALYSIS

WATER 1000 grammes (1 litre).				SOURCE MASSIA.
				Grammes.
Sulphide of sodium	.	.	.	. 0·0142
Carbonate of soda	.	.	.	. 0·0048
Soda .	.	.	.	. 0·0410
Silica .	.	.	.	. 0·0470
Sulphate of soda	.	.	.	. 0·0150
Chloride of sodium	.	.	.	. 0·0140
Lime ⎫ Magnesia ⎬ Sulphate of lime ⎭	.	.	.	. 0·0630
Nitrogenous matter	.	.	.	. 0·0210
Bouis.				0·1600

These waters are drunk, and also used in douches
and in baths; but the latter is the usual mode of administration. They have a great reputation as curative agents for skin diseases, and especially for those of
a herpetic nature; and in this respect M. Bazin, the
eminent dermatologist of the Saint-Louis Hospital,
does not hesitate to place them among the most valuable springs in the chain of the Pyrenees. They are
also useful for chronic rheumatism, painful swellings
of the joints, &c.

The patients can reside either at the Thermal Establishment (where a large drawing-room is placed at
their disposal), or in any of the various comfortable
hotels.

LE VERNET

(*Pyrénées Orientales*)

SODIO-SULPHURETTED WARM SPRINGS

LE VERNET is a village situated about five miles from Prades, at the foot of the peak of Canigou, in a little valley crossed by a bounding torrent known under the name of *Castell*.

This Thermal Station, hardly known a few years ago, deserves the attention of medical men and invalids, from the nature of its waters and the bathing arrangements. It possesses two Establishments, known under the names of *Thermes Mercader* and *Thermes des Commandants*, each including several buildings, some reserved as lodgings for the patients, and others as bath and douche rooms, halls for inhalation and pulverisation, hot-air baths and *buvettes*. The thermo-mineral springs of the *Mercader* establishment are five in number. Two are used for drinking purposes, and two for baths ; while the *Source Ursule* supplies the inhalation and pulverisation rooms. The springs of the Establishment *des Commandants* are seven in number. Their overflow supplies a vast swimming-bath measuring 114 feet by 42 feet, and surrounded by forty very comfortable dressing-rooms.

All these springs, the temperature of which varies from between 18° to 59° C. (64° to 138° F.), have one common origin, and belong to the sodio-sulphuretted

class. The water is clear, tasting and smelling more
or less strongly of sulphur according to the spring,
oily to the touch, and depositing *barégine* in the reservoirs. Their chemical composition is practically
identical, and their degree of mineralisation varies
between gram. 0·2276 and gram. 0·2734 to the litre.

CHEMICAL ANALYSIS

WATER 1000 grammes (1 litre).	SOURCE DE LA PROVIDENCE.
	Grammes.
Sulphide of sodium	0·0420
Sulphite of soda	0·0050
Sulphate of soda	0·0215
,, magnesia	0·0035
,, lime	0·0010
Silicate of soda	0·0628
Carbonate of soda	0·0910
,, potash	0 0100
,, magnesia	0·0020
,, lime	0·0010
Chloride of sodium	0·0160
Alumina	0·0010
Glairine	0·0150
Iodide of potassium . . .	0·0001
	0·2734

Among the affections for which the waters of Le
Vernet may be recommended, stand first : chronic
diseases of the respiratory organs, bronchitis, laryngitis, catarrh, and phthisis in the early stage. Then
come rheumatism, local, general, chronic, or scrofulous;
skin diseases of a herpetic nature ; chronic discharges
of the uterine and vaginal mucous membranes, and of
the male urinary passages. These waters are also

suitable for asthenic gout, constitutional syphilis, and the ill effects of mercury. The special conditions of climate, situation, and general arrangements of Le Vernet render this Station an excellent winter residence for invalids suffering from diseases of the respiratory organs.

The list of interesting excursions which can be made in the neighbourhood includes those to the iron mines of *Torrent d'Escaro*, and to the copper mines of *Canaveilles*, to the ruins of the Monastery of *Saint-Martin*, and lastly, to *Mont Canigou*, 9058 feet high, from which in fine weather a view of the whole range of the Pyrenean mountains can be enjoyed, which it would be difficult to surpass in beauty.—(See Wintering Stations.)

AMÉLIE-LES-BAINS

(*Pyrénées Orientales*)

SODIO-SULPHURETTED WARM SPRINGS

AMÉLIE-LES-BAINS, formerly called indifferently *Arles-les-Bains* and *Bains-sur-Tech*, is a small town of 600 inhabitants, built on the banks of the *Mondoni*, 897 feet above the level of the sea.

This Thermal Station, which dates as far back as the time of the Romans, contains no fewer than three Establishments.

The first and oldest, called *Thermes Romains*, is one

of the most comfortable and best organised in the whole of the Pyrenees; it contains forty-six bath-rooms, seven douches, one vaporarium, one large public bath (*piscine*), two small ones, and in addition complete hydropathic arrangements.

The second and more recent has been built by **Dr. Pujade**, who is still its proprietor and director. It consists of two distinct parts : one is reserved for the use of the waters, and the other is a comfortable hotel or *pension* for the patients. The former comprises thirty-two bath-rooms with marble baths, douches of all kinds, vaporarium, inhalation and pulverisation rooms; besides which there are a large public swimming bath and gymnasium combined, 100 feet square, and ten *burettes*.

The third is *l'Hôpital Militaire*, belonging to the State, and is one of the finest of its kind; it can accommodate nearly 600 soldiers. The bathing part is composed of a public bath of sixty places, six bath-rooms and douches, eight " grand " douches, and one inhalation room, all for the use of the privates. Another public bath for thirty, eight private bath-rooms, and a very varied series of douches, are reserved for the officers.

These three Thermal Establishments are supplied by seventeen springs, which are all sodio-sulphuretted, and of a temperature varying between 20° and 60° C. (68° and 140° F.) All the springs give a clear, colourless water, with a more or less distinct flavour of gall; they deposit *barégine* and confervæ, and become slightly cloudy on contact with the air. It will be enough to give as a type the analysis of one of the springs.

CHEMICAL ANALYSIS

WATER 1000 grammes (1 litre).	SOURCE DU PETIT ESCALDADOU.
	Grammes.
Sulphide of sodium	0·011
Chloride of sodium	0·045
Carbonate of soda and potash . . .	0·087
Sulphate of soda	0·060
Silicate of soda	0·119
Alumina and oxide of iron . . .	0·004
Glairine	0·010
Traces of lime and magnesia—*Poggiale.*	0·336

The waters of Amélie-les-Bains stimulate the digestive functions, excite the nervous system and the circulation, and quickly induce all the particular symptoms of thermal fever, which, however, do not last long. The weaker springs are used for drinking, and the stronger ones are employed for ordinary and vapour baths, douches, and inhalations. They give excellent results in certain cases of chronic bronchial catarrh; in early symptoms of phthisis; in rheumatism, scrofula, caries of bones, white swellings of joints, old sores, and all kinds of wounds.

The climate of Amélie-les-Bains is very mild, so that this little town is an excellent winter resort, and accordingly the baths and hotel establishments are open at all seasons.

N.B.—For promenades and excursions see Wintering Stations.

LA PRESTE

(*Pyrénées Orientales*)

SODIO-SULPHURETTED AND BICARBONATED WARM SPRINGS

LA PRESTE is a little village situated about twenty miles from Amélie-les-Bains. It stands on a narrow plateau which forms a kind of promontory between the two ravines of the Tech and Labane.

The Thermal Establishment has been considerably improved within the last few years ; new buildings have been added for the accommodation of patients, and have all been comfortably fitted up. The number of baths (all of white marble) has been considerably increased, douches of all kinds, and several *buvettes*, have been added.

CHEMICAL ANALYSIS

WATER 1000 grammes (1 litre).	SOURCE APOLLON.
	Grammes.
Carbonate of soda	0·0397
„ potash	traces
„ lime and magnesia . . .	0·0011
Sulphide of sodium . . .	0.0127
Sulphate of soda . . .	0·0206
„ lime	0·0007
Chloride of sodium	0·0014
Silicic acid	0·0421
Barégine or glairine	0·0103
Loss	0 0051
Anglada.	0·1337

The springs, which are all hot, are numerous, but

seem to proceed from one common origin; they have the same chemical composition. The principal, called *Grande Source* or *Source d'Apollon*, is the only one used. It rises within the Establishment, which it supplies, and furnishes 44,000 gallons per day of clear, colourless water, with a slightly sulphurous smell and an alkaline flavour. It deposits white filaments of *glairine* in the reservoirs. Its temperature is 45° C. (113° F.)

It results from this analysis that the waters of La Preste are of a mixed character. They are sodio-sulphuretted and sodio-bicarbonated, and contain an excess of alkaline principles. They are employed for drinking, and for baths and douches. They are easily absorbed by the stomach, increase the appetite, and exercise a remarkable action upon the urinary passages, increasing considerably the secretion of the kidneys, and diminishing the over-acidity of the urine. They are also especially valuable to patients suffering from uric and phosphatic gravel, vesical catarrh, vesical irritability or spasm, and chronic cystitis. In these last cases they are superior to the strong sodio-bicarbonated waters. They facilitate the expulsion of renal calculi; they exercise also a beneficial action in chronic affections of the liver.

In addition to the above, the waters of La Preste will be found beneficial in the cases for which ordinary sulphurous waters are prescribed, such as, for example, bronchitis and chronic laryngitis, *idiopathic* or *spasmodic* asthma, phthisis in the first stage, and certain affections of the skin. The waters of La Preste are exported, and keep well.

Numerous and varied walks can be taken from this Thermal Station, buried as it is in the heart of the Pyrenees. It is chiefly frequented by Spaniards. (See Wintering Stations.)

LE BOULOU

(*Pyrénées Orientales*)

FERRUGINOUS, SODIO-BICARBONATED, AND GASEOUS SPRINGS.

LE BOULOU is situated at about five miles from Céret, and only three from the Spanish frontier.

At a little distance from the village, in the mountain *des Albères*, there are four springs, which may be considered as typical of the ferruginous bicarbonated waters, being richly mineralised, and containing a large quantity of carbonic acid gas. Their temperature varies from 17° to 20° C. (63° to 68° F.)

Le Boulou possesses a small Thermal Establishment, but its waters are chiefly used for exportation, and can be drunk by patients at home. Like all ferruginous alkaline waters, they are suitable for many affections of the digestive organs, dyspepsia, gastralgia, chronic enteritis, and colitis. They are also recommended for certain atonic affections accompanied by relaxation of tissues, such as passive congestion of the liver and spleen, nephritic colic, phosphatic gravel, and chronic gout of the scrofulous diathesis.

SOUTHERN DIVISION

THE mineral springs of the South of France are perhaps more numerous than those of any other Division, but they are far from being fully utilised. Nevertheless, although some among them may be difficult of access, or unprovided with suitable arrangements for the treatment or accommodation of patients, —yet there are others which, by their abundance and their incontestable therapeutical value, have obtained a well-deserved reputation.

Certain Thermal Stations of this region, in consequence of their geographical situation and the special arrangements made at their Establishments, have become at the same time Wintering Stations, and offer to invalids the exceptional advantage of the combination of hydro-mineral treatment with a residence in a favoured climate.

11

BARBOTAN

(*Gers*)

SODIO-SULPHATED AND FERRUGINOUS SPRINGS

Mud-baths.

THE village of *Barbotan* is situated a few miles from *Cazaubon*, in the district of *Condom*. It is built upon piles and artificial foundations, on account of the boggy nature of the soil, which is saturated by its mineral waters. The springs, which are sodio-sulphated and ferruginous, rise to the surface of the ground, and warm the blackish slime which covers it to a temperature varying from 15° to 38° C. (59° to 100° F.) The Barbotan waters have been known from ancient times. The excavations made for the construction of the present Establishment brought to light medals of Roman origin, which proved that, at the time of the invasion of Gaul, the victors founded here a bathing Establishment.

For a long time these waters were administered only in the form of " boues " (mud-baths), and it is greatly to be regretted that no exact analysis has been made of these mineralised deposits. It is, however, known that they contain—1st, the warm springs which are liberated in the depths of the underground natural reservoirs; 2nd, those formed of the overflowings of the springs utilised in the *boues* Establishments; 3rd, a blackish slime carried by the water, which covers the entire surface of the neighbourhood of Barbotan. Consequently the "boues" combine the properties belonging to those thermal waters with the special qualities of the slimy substance. This contains alumina, silica, magnesia, sulphate of lime, and ferruginous oxides.

The Thermal Establishment was renovated a few years ago, and is at present very comfortable. It contains special rooms for baths, douches, "boues," and a well-appointed *buvette*. The waters are transparent, perfectly clear, with a sweetish but somewhat astringent taste, and exhale a faint odour of sulphuretted hydrogen.

CHEMICAL ANALYSIS

WATER 1000 grammes (1 litre).	SOURCE ALEXANDRE.
Carbonic acid . . . . Lit.	0·122
Sulphuretted hydrogen	traces
	Grammes.
Carbonate of lime	0·0210
„ magnesia	0·0020
„ iron	0·0312
Sulphate of soda	0·0312
„ lime	0·0020
Chlorides of sodium and magnesium . .	0·0190
Silicic acid and barégine	0·0290
	0·1354

The Barbotan waters are much frequented by invalids from the neighbouring Departments. The diseases against which they act most efficaciously are— rheumatism in all its various and complicated manifestations; neuralgia, principally that which affects the great plexuses and the chief nervous trunks (lumbar, sciatic, intercostal neuralgia, &c.) ; local paralysis, and those cases which are independent of any acute or recent lesion of the nervous centres; locomotor ataxy, and affections belonging to the herpetic, scrofulous, ricketty, and syphilitic diatheses.

They are not suited to persons liable to cerebral congestion or gout, that is to say, to those of plethoric constitutions.

Life in Barbotan is very quiet, and the only amusements to be had are reading, games, and walking in the neighbourhood.

CASTERA-VERDUZAN

(Gers)

CALCAREO-SULPHURETTED SPRINGS, AND FERRUGINOUS SPRING.

CASTERA-VERDUZAN is a village containing 1000 inhabitants, situated in a pretty valley, about seventy-five miles from Bordeaux, and nearly nine from *Auch*.

Its mineral waters are calcareo-sulphuretted and ferruginous. The sulphurous waters flow from two springs, the *Source Grande-Fontaine* and *Source Petite Fontaine,* yielding altogether more than 66,000 gallons daily of water at a temperature of 23·5° C. (74° F.)

The ferruginous water proceeds from only one spring, which is cold.

At Castera, as at Cambo, patients have the great advantage of being able to combine sulphurous with ferruginous treatment.

The following is the analysis of the sulphureous spring:

CHEMICAL ANALYSIS

WATER 1000 grammes (1 litre).	Grammes.
Sulphuretted hydrogen	0·0003
Sulphide of calcium	0·0006
Sulphate of soda	0·1070
,, lime	0·5163
,, magnesia	0·2420
Carbonate of lime	0·2300
,, magnesia	0·1920
Chloride of sodium	0·0302
Oxide of iron	0·0015
Silica	0·0130
Organic matter	0·0180
Ammonia	0·0018
Traces of sulphate of potash, borate of soda, and iodine.—*Filhol.*	1·3527

The Establishment is very comfortable and elegant, and includes a *buvette,* thirty bath-rooms, and apartments for douches.

Among the affections most often to be found at this station, may be mentioned chlorosis, anæmia, leucorrhœa, gastralgia, and chronic gastritis; bronchial and pulmonary catarrh; hysterical and hypochondriacal affections; and lastly, skin diseases and rheumatism.

DAX

(*Landes*)

HOT ALKALINE SPRINGS AND MUD-BATHS

THE little town of *Dax* occupies the middle point of a triangle, of which Arcachon, Pau, and Biarritz form the angles; and the great railway which con-

nects Paris with Madrid runs along its walls. Its
waters were known in the time of the Romans, under
the name of *Aquæ Tarbellicæ.* The spring which
rises in the very heart of the town, bears the name
of *Fontaine chaude* or *Fontaine de Nesle.* Its temperature is 60° C. (140° F.) ; it is inodorous, tasteless,
and perfectly clear, and yields from 2,187,500 to
2,625,000 gallons every twenty-four hours.

CHEMICAL ANALYSIS

WATER 1000 grammes (1 litre).	SOURCE FONTAINE CHAUDE.
	Grammes.
Carbonate of magnesia . • • .	0·027
Sulphate of soda . • • .	0·151
„ lime . • • .	0·170
Chloride of sodium . • • .	0·032
„ magnesium . • • .	0·096
Thore and Meyrac.	0·475

In a new analysis, M. Meyrac has found traces of
alkaline iodides and bromides in the confervæ of this
spring.

The Thermal Establishment stands just where the
Sainte Marguerite and *du Bastian* springs issue from
the ground, and is built upon a bank of mineralised
mud, formed by the deposits of the river Adour. Its
arrangements are of the most complete and comfortable description. Patients can live either in the
Establishment or in the town.

Mineralised mud-baths constitute the characteristic
treatment of Dax. In 1000 grammes of solid matter
they contain 800 grammes of silica, and 200 grammes

of vegeto-mineral substances, principally alumina, oxide and proto-sulphuret of iron, magnesia, chloride of sodium, and organic matter. The "muds" are used in large public baths or *piscines*, through which new currents of mineral water flow continually, keeping the baths at the most suitable temperature. The heat of these baths varies (according to the indication) from 30° to 45° C. (86° to 113° F.), and can even be raised to 49° and 50° C. (120° and 122° F.) The mineralised mud is also used in private baths, both general and local. This Establishment may be called a model one, as regards the therapeutical value of the water supply, the general arrangements, and the excellent management.

In consequence of these various modes of administration, and the constant and enlightened supervision exercised by the two physicians residing at the Establishment, the Thermes of Dax have considerably extended the sphere of application of its mud-baths and mineral springs.

The following list taken from the latest clinical returns shows the diseases for which they are best suited:—Rheumatism and chronic gout; neuralgia and nervous affections; paralysis (paraplegia, hysterical and rheumatic paralysis); stiffness and enlargements of the joints, consequent on rheumatism; uterine affections; scrofula and skin diseases.

These baths are not only frequented during the ordinary summer season, but also afford all the advantages of a winter residence. (See Wintering Stations.)

CAMBO

(Basses-Pyrénées)

CALCAREO-SULPHURETTED AND FERRUGINOUS SPRINGS

CAMBO is a little town built on the river *Nive*, 160 feet above the level of the sea, and only nine miles from Bayonne.

The calcareo-sulphuretted waters proceed from only one spring, which yields no less than 220,000 gallons in twenty-four hours, at a temperature of 23° C. (73° F.)

Cambo also possesses a ferruginous spring of a temperature varying from 15° to 16° C. (59° to 61° F.) Its water is clear and limpid when it issues from the ground, but loses its transparency on contact with the air; yellow flakes form within it, and it becomes covered with an iridescent film.

The Thermal Establishment stands on the left bank of the river *Nive*, not quite a mile from the town, and contains several bath-rooms, douches, and two *buvettes.*

SALIES-DE-BÉARN

(Basses-Pyrénées)

COLD SODIO-CHLORINATED SPRINGS

SALIES-DE-BÉARN is a pretty little town of 9000 inhabitants, situated nearly ten miles from *Orthez*, at the end of a delightful valley. If tradition may be believed, the salt springs were discovered accidentally ; and soon there gathered around them a numerous population, which grew important enough to form first a village, and afterwards a town, called *Salies* (salt town).

The sodio-chlorinated waters of Salies have their different sources at the base of a hill partly composed of gypsum, and meet in a reservoir in the middle of the town. This large tank remained for a long time exposed to the air, but has now been covered over. It measures about fifty feet each way, and supplies an extensive manufactory where the salt is prepared. There the water is heated in large shallow iron boilers ; the evaporation is carried on by an open fire, kept up night and day without intermission, and sends into the air vapours containing appreciable quantities of chloride of sodium. The inhabitants thus find themselves under nearly the same atmospheric conditions as if they were living on the sea-shore, except that they are not exposed to violent winds, which often render such a residence so trying.

The analysis of the condensed waters, *i.e.* of those left after evaporation in the boilers of the salt manufactory, has given fifteen centigrammes of alkaline iodides (besides the chlorides) to the litre or 1000 grammes. This greater proportion of iodides in the condensed water agrees perfectly with all that is known of the facility with which iodide of sodium is precipitated during the ebullition of the liquid which

contains it. The Salies waters are clear and colourless, tasting strongly of salt, with a bitter after-taste;
their density is 1208. With regard to the *eaux mères*
or condensed waters, they are in general without
smell, of a tawny or brownish colour, and an acrid
and saline flavour; their density is also very high.
Comparatively speaking, those of Salies are less deeply
coloured than most, their taste is the same, and their
density is equal to 1221.

The following is the analysis of the natural mineral
water of the spring :

CHEMICAL ANALYSIS

WATER 1000 grammes (1 litre).	
	Grammes.
Chloride of sodium	216·020
,, potassium	2·080
Sulphate of potash ⎫	
,, soda ⎬	9·750
,, magnesia ⎪	
,, lime ⎭	
Alkaline bromide	1·050
Phosphates, silica, alumina, bicarbonates of lime and magnesia	5·500
Traces of chloride of calcium and magnesium, alkaline iodide, oxide of iron, and organic matter.— *Reveil and O. Henry.*	233·400

The physiological effects of the Salies waters vary
according to their different degrees of temperature,
or mineralisation, and the methods of administering
them,—whether as drink, baths, or cold, temperate, or
hot douches.

The waters are taken internally in two different ways:

1st, cold, pure, and in large quantities; 2nd, hot, mixed with chicken broth, and in small glasses containing four ounces. It is to be noticed that they act as purgatives when taken cold; but, on the contrary, when taken warm they cause no purging, and act as dissolvents, increasing all the secretions, exciting the appetite, facilitating digestion, and thus, by inducing a quantitative improvement of the nutrition, cause a general physiological improvement. Taken in baths at a temperature of 23° C. (82° F.), they are soothing, sedative, tonic, resolutive, and restorative, owing to the cutaneous absorption which quickly takes place. If used in baths at a higher temperature than 30° C. (86° F.), they become very exciting, increasing the activity of the circulation, congesting the skin, which loses its absorbing power,—inducing profuse perspiration of the face, accompanied by headache,—and considerably diminishing the secretion of urine.

It is easy to deduce from the composition of the Salies waters the class of diseases for which they may be used with the greatest success. In the first place stand chlorosis and anæmia; scrofula and its different manifestations, whether ganglionic, cutaneous, or osseous. In the second place may be mentioned rheumatism and all its consequences; certain cases of paralysis, locomotor ataxy and muscular atrophy; displacement of the uterus, chronic metritis, either with or without catarrh, as well as amenorrhœa and dysmenorrhœa.

RENNES-LES-BAINS

(*Aude*)

FERRUGINOUS BICARBONATED SPRINGS, AND SODIO-CHLORINATED SPRINGS.

RENNES-LES-BAINS is a village situated on the banks of the river *Salz*, in a ravine between mountains of moderate height. It appears that the Romans were well acquainted with its mineral waters, for excavations in the village have led to the discovery of medals, coins, remains of ancient buildings, and fragments of inscriptions.

There are five springs, named respectively :—*Source du Bain fort*, of which the temperature is 51° C. (124° F.) ; *Source du Bain doux*, 40° C. (104° F.) ; *Source du Bain de la Reine*, 31° C. (88° F.) ; *Source de l'Eau du pont*, and *Source de l'Eau du cercle*, both 12° C. (54° F.)

The Thermal Establishment is very well arranged ; it comprises more than thirty bath-rooms, several apartments for different kinds of douches, and a *buvette*.

Besides the five springs just mentioned, Rennes possesses a sodio-chlorinated spring, or rather river, for it is the *Salz* itself. M. Ossian Henry proved by an analysis made in 1839, that the waters of this stream contain chloride of sodium, and it now (as well as the smaller springs) supplies the Thermal Establishment.

CHEMICAL ANALYSIS

WATER 1000 grammes (1 litre).		SOURCE DU BAIN FORT.
Carbonic acid . . . . Lit.		0·162
Sulphuretted hydrogen . . . ,,		0·162
		Grammes.
Carbonate of lime		0·250
,, magnesia		0·070
Chloride of sodium		0·071
,, magnesium		0·280
Sulphate of soda and magnesia . . .		0·090
,, lime		0·162
,, iron		0·162
Silicic acid, alumina, phosphate of lime or alumina .		0·049
Carbonated oxide of iron		0·031
Organic matter		0·040
Traces of chloride of potassium and manganese.— *O. Henry.*		1·205

From the preceding analysis, the difficulty of classifying the Rennes waters is sufficiently apparent. They contain chlorides, bicarbonates, and sulphates, in almost equal proportions, and also a considerable quantity of iron. Hydrologists, however, place them among ferruginous bicarbonated waters, and this classification may properly be adopted here.

The waters are administered in baths and douches, and taken as drinks, either pure or mixed, according to the nature of the cases under treatment.

It is easy to understand what advantages can be derived from the united or separate use of these ferruginous and sodio-chlorinated waters, the former being exciting, diuretic, and tonic, and the latter purgative, restorative, and resolutive.

They have a remarkable therapeutical effect on rheu-

matism, neuralgia, uterine neuralgia, spasmodic amenorrhœa, leucorrhœa, chlorosis and anæmia; *lymphatism*, glandular congestions of a scrofulous nature,
white swellings, and false anchylosis of joints, &c.

The waters of Rennes-les-Bains are very much
frequented.

ALET

(*Aude*)

CALCAREO-BICARBONATED WARM SPRINGS, AND FERRUGINOUS SPRING.

ALET is a little town of 2000 inhabitants, situated in
a charming valley, 650 feet above the level of the sea,
and surrounded by mountains about 3280 feet high.

It contains a Thermal Establishment, which has of
late years been much enlarged and improved.

The waters are of two kinds, the first belonging to
the alkaline class and being calcareo-bicarbonated, the
second being ferruginous. They proceed from four
springs, three of the former (which yield together
131,250 gallons per day), and one of the latter. It will
be sufficient to give the analysis of the principal alkaline spring, called the *Source des Bains*, of which the
temperature is 28° C. (82° F.) The ferruginous spring
has a temperature of from 10° to 11° C. (40° to 52° F.)

CHEMICAL ANALYSIS

WATER 1000 grammes (1 litre).	SOURCE DES BAINS.
	Grammes.
Bicarbonate of lime ,, magnesia } . . .	0·287

Sulphate of soda
  ,,   lime } .   .   .
  ,,   magnesia
	0·068

Chloride of sodium } Salts of potash } . . .	0·052
Phosphates, insoluble } ,, soluble } . . .	0·080
Silica, alumina } Organic matter, iron, and loss } . .	0·040
O. Henry.	0·527

The waters of Alet are used as drinks, and for baths and douches. They have been successfully prescribed for sick headaches, chlorosis, and what is commonly called nervousness. They are tonic, and useful in cases of general debility.

These mineral waters are exported largely, and keep without undergoing any deterioration.

During the course of treatment patients can enjoy many interesting excursions in the neighbourhood of Alet. The peak of *Bugarach*, which overlooks all the Roussillon county ; the grottos of *Fos ;* the *Pierre Lis,* a road traced in a sort of groove in the rock; the intermittent fountain of *Belesta,* the ruins of the church of *Nôtre Dame d'Alet,* the *Château d'Arques,* &c.

CAMPAGNE

(*Aude*)

CALCAREO-BICARBONATED WARM SPRINGS

CAMPAGNE is a small town situated on the left bank
of the *Aude*, about four miles from Alet, and nearly
two miles from Esperrazza, where patients are obliged
to lodge and live. Its waters, which are calcareo-
bicarbonated, were formerly very celebrated, and the
annual number of visitors has been known to amount
to 3000. There are three springs, yielding nearly
110,000 gallons of water per day. The two principal
are the *Source des Bains*, of which the temperature
rises to 31° C. (88° F.), and which is employed in baths
and douches, and the *Source de la Buvette*, of which
the temperature only reaches 29° C. (84° F.) This
latter, as its name implies, is exclusively used for
drinking.

The Campagne waters are sweet, clear, and colourless, with a characteristic smell, and a taste which,
without being styptic, is slightly ferruginous. When
they first emerge from the ground, bubbles rise in
considerable numbers, and burst on their surface.

Free gases.					Litre.
Carbonic acid	.	.	.	.	0·108
Oxygen	.	.	.	.	0·002
Nitrogen	.	.	.	.	0·020

CHEMICAL ANALYSIS

WATER 1000 grammes (1 litre).	SOURCE DE LA BUVETTE.
	Grammes.
Carbonate of lime	0·346
„ magnesia	0·032
Sulphate of lime	0·058
„ soda	0·084
„ magnesia	0·170
„ potash	0·019
Chloride of sodium	0·035
„ potassium	0·012
Silica	0·020
Carbonated and crenated oxide of iron .	0·005
Organic matter	0·032
Traces of chloride of magnesium, oxide of manganese, fluoride of calcium, arsenic, and iodine	0·813

The Campagne waters are beneficial in cases of vesical catarrh and gravel. They may also be prescribed for atonic affections of the stomach and intestinal canal, chronic congestion of the liver, leucorrhœa, chlorosis, and anæmia. Some authors recommend them very highly, not only for the results of intermittent fevers, but for the fevers themselves.

LA MALOU

(*Hérault*)

FERRUGINOUS BICARBONATED WARM SPRINGS

The springs of La Malou rise in a pleasant valley, enclosed by mountains covered to their very summit with vines and chestnut trees. They must be classed as ferruginous, alkaline waters; but they possess the exceptional character of having a high degree of thermality.

The different springs are all nearly identical in composition and physical characteristics, and have a temperature varying from 16° to 36° C. (61° to 97° F.) The water is clear, of a decidedly inky flavour, with an acidulous after-taste. The springs have been conducted into three Establishments, placed at a trifling distance apart, and known under the names of *La Malou-le-Bas*, *La Malou-le-Centre*, and *La Malou-le-Haut*. *La Malou-le-Centre* has only recently been rebuilt; it is decidedly the most important and best organised of the three.

CHEMICAL ANALYSIS

WATER 1000 grammes (1 litre).					SOURCE LA MALOU LE BAS.
Carbonic acid .	.	.	.	Lit.	0·8280
					Grammes
Bicarbonate of soda	.	.	.	.	0·7711
„ potash	.	.	.	.	0·1242
Carbonate of lime	.	.	.	.	0·4528
„ manganese .	.	.	.	.	0·1863
Peroxide of iron	.	.	.	.	0·0251
Chloride of sodium	.	.	.	.	0·0187
Silicic acid .	.	.	.	.	0·0638
Alumina .	.	.	.	.	0·0302
					1·6722

The waters of La Malou are drunk, but are chiefly used for baths, douches, and inhalations. The external treatment is mostly carried on in *piscines* or public baths, filled by running water.

These waters are beneficial for all the varieties of chronic rheumatism, but particularly in cases of rheumatoid arthritis, accompanied with nodosity of the joints; also for chlorosis, anæmia, and uterine affections caused by poverty of blood. Their speciality, however, consists in the treatment of nervous diseases. Cures have often been effected in inveterate cases of neuralgia and neurosis; also paralysis of different kinds, but especially those resulting from disease of the spinal cord, of which imperfect motion of the lower limbs, and a marked deterioration of general health, are the early and invariable signs. These waters have also acquired a reputation in cases of sexual impotence.

AVESNE

(*Hérault*)

WARM ALKALINE GASEOUS SPRINGS

AVESNE is a village situated ten miles from *Lodève*, and fifty from *Montpellier*, at a height of 933 feet above the level of the sea. The springs contain bicarbonates of soda and lime, they are very abundant, and their temperature is 28° C. (82·4° F.)

CHEMICAL ANALYSIS

WATER 1000 grammes (1 litre).				Grammes.
Carbonate of soda	.	.	.	0·1028
,, lime	.	.	.	0·0995
Sulphate of magnesia	.	.	.	0·0687
Chloride of sodium	.	.	.	0·0462
Silicic acid	.	.	.	0·0045
Alumina	.	.	.	0·0062
Oxide of iron	.	.	.	traces
				0·3279

The waters of Avesne are considered both tonic and sedative, and are prescribed for cutaneous affections, particularly of the humid, crustaceous, and pustulous types.

BALARUC

(*Hérault*)

SODIO-CHLORINATED WARM SPRING

BALARUC is a little village situated on the banks of the lake of *Thau*, an inlet of the Mediterranean sea, opposite Cette, at a short distance from Montpellier, and about two miles from *Frontignan*, so famous for its wines. The mineral spring to which Balaruc owes its present celebrity, was much patronised by the Romans, as is proved by the magnificent white marble *piscina* lately discovered near the source. The waters of this Station contain strong proportions of chloride of sodium; they are inodorous, but their clearness is disturbed by reddish-grey pellicles, which are held in suspension, and collect on the surface. They proceed from only one spring, which yields no less than 66,000 gallons daily, at a temperature of 48° C. (118° F.) Authors believe that this spring has a submarine origin; for they have observed that the proximity of the lake of Thau (which is only at a few miles' distance from the sea, and communicates with it,) exercises a certain influence upon the variations in their temperature.

CHEMICAL ANALYSIS

WATER 1000 grammes (1 litre).	Grammes.
Chloride of sodium	6·802
„ magnesium	1·074
Sulphate of lime	0·803
„ potash	0·053
Carbonate of lime	0·270
„ magnesia	0·030
Silicate of soda	0·013
Bromide of sodium	0·003
„ magnesium	0·032
Oxide of iron	traces.
MM. de Serres and L. Figuier.	9·080

The waters of Balaruc belong to the same class as those of Bourbonne-les-Bains and Wiesbaden; they are laxative in doses of four or five glasses; but when taken in larger quantities, become thoroughly purgative. When administered in baths at a medium temperature, they are very strengthening, and have a valuable influence upon disorders of the nervous system; but when taken at an elevated temperature, they produce a strong revulsive action. They are principally recommended for cases of partial paralysis of a chronic form; indeed, their remarkable efficacy in the treatment of these special affections has obtained for them a European reputation. They are also suitable for various other diseases, such as chronic rheumatism, sciatica, gunshot wounds, false anchylosis, white swellings, caries, necrosis, &c. Lastly, they may be prescribed with benefit for the scrofulous diathesis, which they considerably modify.

The best time for visiting Balaruc is during the months of May, June, September, and October, the heat being too overpowering in July and August. The only recreations to be had at this quiet Station are walks on the banks of the lake, or excursions and drives by the sea-shore.

AIX

(*Bouches-du-Rhône*)

CALCAREO-BICARBONATED WARM SPRINGS

THE town of *Aix* is now more celebrated for its oils than for its mineral waters, although it was the reputation of its Thermal Springs which had made it—before the Christian era—one of the most flourishing cities of Gaul.

The principal spring, called *Source de Sextius*, is now received in a small Establishment built by the side of the ancient Roman *Thermæ*,—the great *piscinæ* belonging to which, now unused, are for the most part buried under the earth. The water of this spring is perfectly clear, tasteless, and inodorous, with a temperature of from 34° to 36° C. (93° to 97° F.) It is at first slightly stimulating, but has ultimately a sedative effect.

CHEMICAL ANALYSIS

WATER 1000 grammes (1 litre).	SOURCE SEXTIUS.	SOURCE BARRET.
	Grammes.	Grammes.
Carbonate of lime . . .	0·1072	0·2416
„ magnesia . .	0·0418	0·1080
Chloride of sodium . . .	0·0073	0·0070
„ magnesium . .	0·0120	0·0286
Sulphate of soda . . .	0·0365	0·0880
„ magnesia . .	0·0080	0·0230
Silicic acid and organic matter . .	0·0170	0·0214
	0·2258	0·5176
Traces of iron and carbonic acid.—*Robiquet.*		

The waters of Aix, which are principally used in baths, do not attract many patients from the regions lying beyond Provence. In the opinion of invalids, and even of many medical men, these waters possess rather a hygienic than a curative action, and all their efficacy consists in gently stimulating the system.

GRÉOULX

(*Basses-Alpes*)

SULPHURETTED WARM SPRINGS

THE springs of *Gréoulx* are situated about a mile from the village of the same name, upon the southern slope of the *Alpes-Orientales*. As their temperature never varies from 36° C. (97° F.), they can be employed directly in the baths, and their great abundance allows of a constant flow; while, in addition to these advantages, their sulphurisation does not seem to be inferior to that of the Pyrenean springs.

These waters are beneficial in most affections, whether internal or external, which are recognised as falling within the sphere of sulphurous treatment.

At Gréoulx there is a moderately large Thermal Establishment, where invalids can reside while undergoing their cure.

MONTMIRAIL

(*Vaucluse*)

PURGATIVE WATER (COLD SPRINGS CONTAINING SULPHATES OF MAGNESIA AND SODA)—SULPHUROUS SPRING.

MONTMIRAIL is a little village in the district of Vaucluse, situated between *Orange* and *Carpentras*, and possessing two mineral springs, one of which is sulphurous, and the other contains sulphates of soda and magnesia.

The latter of the two is by far the more interesting, as being unique of its kind in France. It is destined to be very extensively patronised by patients when better known, and is expected to supersede the celebrated German purgative waters of Seidlitz and Pullna. Like them, it contains a considerable quantity of sulphate of magnesia, as has been proved by the analysis of M. Ossian Henry.

CHEMICAL ANALYSIS

WATER 1000 grammes (1 litre).	Grammes.
Sulphate of magnesia	9·31
„ soda	5·06
„ lime	1·00
Chloride of magnesium	0·83
„ sodium ⎱	0·18
„ calcium ⎰	
Bicarbonate of lime	0·37
„ magnesia	0·16
Calcareous phosphates, silica, alumina, sesquioxide of iron, and arsenical principle	0·39
Traces of iodides, salts of potash and ammonia, and organic matter.—*O. Henry.*	17·30

The water of this spring, also called *Source Verte* on account of its colour, has a bitter flavour, owing to the sulphate of magnesia that it contains, and possesses the same purgative properties as the waters of Pullna and Seidlitz. One bottle is sufficient fully to produce the desired effect, which begins half an hour after taking it.

One tumblerful acts as a gentle laxative, and is not followed afterwards by constipation.

According to M. Poudet, it has two important advantages over its German rivals :—(1) It has a much less disagreeable taste ; (2) It purges without causing any intestinal irritation or griping. The waters of Montmirail can therefore be advantageously used in every case where hitherto those of Seidlitz or Pullna have been employed,—as, for instance, indigestion and constipation, abdominal congestion and obstruction, affections of the liver and spleen, sequelæ of intermittent fevers, &c.

The purgative water of Montmirail keeps perfectly well, and can be exported without undergoing any deterioration.

BONDONNEAU

(Drôme)

CALCAREO-BICARBONATED COLD SPRINGS

BONDONNEAU is a large village situated only two miles from *Montélimar*, and about three from the Rhône, upon a large plateau overlooking a magnificent plain.

Its cold calcareo-bicarbonated waters were discovered in 1854. They proceed from three springs, yielding together nearly 22,000 gallons in the twenty-four hours. They are clear, and at their source have a somewhat sulphurous smell; and they allow the escape of carbonic acid gas, of which they contain an excessive quantity. They strongly resemble the Saint-Galmier and Condillac waters.

CHEMICAL ANALYSIS

WATER 1000 grammes (1 litre).	
Sulphuretted hydrogen Very perceptible at the source	
Carbonic acid $\frac{3}{4}$ of the volume of the water	
	Grammes.
Bicarbonate of lime ⎫ „ magnesia ⎭ . . .	0·330
„ soda	0·006

Sulphate of soda   ⎫
     „   lime   ⎬   .   .   .
     „   magnesia ⎭
	0·043

Chloride of sodium	0·030
Alkaline bromides and iodides . . .	0·003
Sesquioxide of iron with manganese . .	0·002
Silica and alumina	0·128
Traces of salts of potash, arsenical principle, phosphates, and nitrogenous organic matter.—*O. Henry.*	0·602

The Thermal Establishment, which is of recent construction, includes twenty-five well-appointed bathrooms, as well as rooms for douches and inhalations, and a hot-air room.

The Bondonneau waters are strengthening, stimulating, and digestive, and increase the urinary and cutaneous secretions. Under their influence, the circulation becomes more active, and there is a marked increase in the number and force of the arterial pulsations. They are, in consequence, absolutely inadmissible for patients predisposed to congestion of any important organ, such as the brain, the lungs, or the liver ; and for those who are affected with organic disease of the heart. On the other hand, they may be recommended with great advantage to patients suffering from chronic affections of the digestive organs, such as dyspepsia and chronic diarrhœa, or those subject to laryngitis, bronchitis, and chronic pleurisy. Anæmic and chlorotic persons will also derive great benefit from taking these waters, which are rich in iron, arsenic, and iodine, as shown by the analysis above quoted.

CONDILLAC

(Drôme)

GASEOUS, CALCAREO-BICARBONATED COLD SPRINGS.

CONDILLAC is a village in the neighbourhood of Montélimar, the waters of which have become largely used during the last few years. They had been known to the Romans, but were afterwards forgotten, and only rediscovered in the year 1845, beneath the rocks and earth fallen for centuries from *Mont Givode.* They are calcareo-bicarbonated, and come from two springs called *Source Anastasie* and *Source Lise.*

These two springs, of which the *Source Anastasie* is the more important, supply daily more than 6600 gallons of water, which is clear, colourless, inodorous, sharp to the taste, and sparkling in the glass. Numerous bubbles of gas escape from its surface, especially at the point where it emerges from the ground.

CHEMICAL ANALYSIS

WATER 1000 grammes (1 litre).			SOURCE ANASTASIE.
Carbonic acid, free ⎫ Sulphuretted hydrogen ⎭	.	. Lit.	0·548
			Grammes.
Bicarbonate of lime	. . .	.	1·359
„ magnesia	. . .	.	0·035
„ soda	. . .	.	0·166
Sulphate of soda	. . .	.	0·175
„ lime	. . .	.	0·053
Chloride of sodium	. . .	.	0·150
„ calcium	. . .	.	0·150
Silicate of lime and alumina	. .	.	0·245
Carbonate and crenate of iron	. .	.	0·010
Traces of salts of potash, nitrates, iodides, and organic matter.—*O. Henry.*			2·193

The Condillac, like the Saint Galmier, are chiefly table-waters, and are excellent for dyspepsia, gastralgia, chronic gastritis, flatulence, sour eructations, heart-burn, &c. They have been successfully taken to check the distressing vomiting during pregnancy; and are also efficacious in chlorosis and anæmia. Condillac itself is little visited, but its waters are largely exported.

CELLES

(*Ardèche*)

CALCAREO-BICARBONATED COLD SPRINGS, AND FERRUGINOUS SPRINGS.

CELLES is quite a small hamlet, situated three miles from *Lavoulte ;* its mineral waters rise in a valley on the right bank of the Rhône, and proceed from eight springs, of which four contain bicarbonate of lime, and four are ferruginous.

The *Source Bonne Fontaine* is the most important of the former group. Its clear cold water flows at the rate of three gallons a minute ; and its surface is covered by a pearly, iridescent film, whilst the bottom of the basin is lined with a thick red ochreous sediment.

Next in importance comes the *Source du Puits-artésien,* which, although formerly continuous, has now become intermittent. When flowing it gives daily 22,000 gallons of water at 25° C. (77° F.), and more than 120 cubic feet of pure carbonic acid gas, which is collected in a gasometer.

All these springs are utilised in a very comfortable Establishment, which contains bath-rooms, vapour douches, ascending and descending douches, and a vaporarium.

The Celles waters are administered as drinks, and in baths, douches, and inhalations. The natural carbonic acid gas evolved from the mineral springs is also used for air baths and inhalations. It is stated that cancerous, scrofulous, and consumptive diseases are among the affections cured at Celles ; and this assertion is confirmed and supported by observations of Dr Saint-Ange Barbier, made particularly with regard to the cure of cancer, and detailed in a recent work. It is, however, far from being proved that the results actually obtained have equalled the sanguine expectations entertained.

VALS

(Ardèche)

SODIO-BICARBONATED, GASEOUS, COLD SPRINGS,—AND FERRUGINOUS, ALKALINE, AND ARSENICAL SPRINGS.

VALS is a pretty little town situated about one mile and a half from *Aubenas*, in the midst of the extinct volcanoes of the *Vivarais*, and standing in a narrow valley watered by the river *Volane*, one of the tributaries of the Ardèche. It possesses two Thermal Establishments.

The larger and better arranged of the two is supplied with water from the springs so well known by the names of *Source Rigolette, Source Précieuse, Source Madeleine, Source Désirée, Source Saint-Jean* and *Source Dominique.*

The other Establishment receives its water supply

from the *Source Marquise, Source Souveraine, Source Pauline, Source Cloé, Source des Convalescents, Source Saint-Louis,* and *Source Constantine.*

This Station thus possesses fourteen important springs, all of which rise on the left bank of the Volane ; the strata from which they spring consisting of granite, gneiss, and quartziferous porphyry. They may be classified in two groups :

1. *Sodio-bicarbonated springs,* comprising the Rigolette, Précieuse, Madeleine, Désirée, and Saint-Jean, of the first Establishment; and the Marquise, Souveraine, Pauline, Cloé, des Convalescents, and Constantine, of the second.

2. *Ferruginous springs* (containing alkaline sulphates and arsenic), comprising the Dominique, belonging to the first, and the Saint-Louis, belonging to the second, Establishment.

Thus each Establishment has several sodio-bicarbonated springs, and one ferruginous, alkaline, and arsenical spring. We will commence with the first class, as being the most important.

The Vals sodio-bicarbonated springs are cold, their temperature, which is constant in each one, varying between 13° and 16° C. (55° and 60° F.) The water is clear, limpid, and soft in the bath ; it has a peculiarly characteristic smell, due to the presence of a large quantity of carbonic acid,—and also a very pronounced piquant and acidulous flavour, with an alkaline after-taste.

CHEMICAL ANALYSIS

Springs of the First Establishment

WATER 1000 grammes (1 litre).	MADE-LEINE.	DÉSIRÉE	PRÉ-CIEUSE.	RIGO-LETTE.	ST. JEAN
Carbonic acid, free . . Lit.	2·050	2·145	2·218	2·095	2·425
	Gram.	Gram.	Gram.	Gram.	Gram,
Bicarbonate of soda . . .	7·280	6·040	5·940	5·800	1·480
,, potash . . .	0·255	0·263	0·230	0·263	0·040
,, lime . . .	0·520	0·571	0·630	0·259	0·310
,, magnesia .	0·672	0·900	0·750	0·259	0·120
,, iron and manganese	0·029	0·010	0·010	0·024	0·006
Chloride of sodium . . .	0·160	1·100	1·080	1·200	0·060
Sulphate of soda, lime . .	0·255	0·200	0·185	0·220	0·054
Silica, silicates, and alumina .	0·097	0·058	0·060	0·060	0·080
	9·268	9·142	8·885	8·085	2·151

Traces of alkaline iodides, arsenic or arseniates, bicarbonate of lithia, and organic matter.—*O. Henry.*

As will be seen from the preceding analysis, the composition of the waters of Vals approaches that of the Vichy waters; but the former are certainly richer in bicarbonate of soda than any other known springs, and the same may be said with regard to carbonic acid. As a matter of fact, we find the bicarbonate of soda varying per litre from gram. 2·151 (Saint Jean) to gram. 9·268 (Madeleine), and the carbonic acid from 1·082 (Pauline) to 2·500 (Marquise).

These springs (the sodio-bicarbonated) may be divided, according to their chemical composition, into two classes :—1. Those strong in bicarbonate of soda; 2. Those with a moderate or small proportion of the same salt.

CHEMICAL ANALYSIS

Springs of the Second Establishment

WATER 1000 grammes (1 litre).		MARQUISE.	CONSTANTINE.	SOUVERAINE.	CLOÉ.	CONVALES-CENTS.	PAULINE.
Carbonic acid, free	Lit.	2·500	2·1000	2·2000	1·626	1·2400	1·0820
		Grammes.	Grammes.	Grammes.	Grammes.	Grammes.	Grammes.
Bicarbonate of soda .		7·154	7·0530	6·5150	5·289	1·7140	1·6117
,, potash .		—	0·0710	0·0690	0·045	traces	traces
,, lime .		0·180	0·4370	0·2700	0·169	0·0538	0·0288
,, magnesia		0·125	traces	0·0090	0·166	traces	0·0083
,, iron .		0·015	0·0067	0·0056	0·021	0·0475	0·0090
Chloride of sodium .		0·060	0·2800	0·3370	0·189	0·2280	0·0414
Sulphate of soda } ,, lime }		0·053	0·0204	0·2610	0·173	0·4270	0·1696
Silicates and silica .		0·116	—	0·1020	0·099	0·1390	0·1824
Alumina and phosphate of iron					0·001		
		7·703	7·8681	7·5686	6·152	2·6093	2·0512
Traces of bicarbonate of lithia, organic matter, alkaline iodides, and arsenic or arseniates.							
Authors of the Analyses......		Berthier.	O. Henry & Lavigne.	O. Henry & Lavigne.	Dupasquier.	O. Henry & Lavigne.	O. Henry & Lavigne.

The first class (Madeleine, Marquise, Rigolette, Désirée, &c.) are pre-eminently stimulating, resolutive, and restorative. "When drunk," said Palissier in his report to the Academy in 1854, " they increase the appetite, facilitate digestion, regulate the evacuations of the intestines, and at times have a purgative effect. Under their influence the skin becomes warmer, the circulation more active, and an unusual sensation of strength and comfort is experienced. A few glasses are sufficient to render alkaline—perspiration and urine which are naturally acid."

The springs which come under the heading "moderately mineralised" (Marie, Pauline, Saint Jean, &c.) possess, but in a much lesser degree, most of the properties of the more powerful ones. It has been proved that they have in all cases a much more definite action on the urinary organs than the former, their diuretic properties being incontestable. In addition, their stimulating effect is less violent, and they agree much better with some weak and irritable stomachs.

It has already been said that—besides the sodio-bicarbonated springs—Vals possesses others of the ferruginous, alkaline, and arsenical class. These springs, of which there are two (*Dominique* and *Saint-Louis*), have no analogy to the soda springs, and form a distinct group.

The *Source Dominique*, which was named at the commencement of the seventeenth century by a Dominican monk, who was the first to make use of it and test its beneficial effects, rises in the midst of the other springs, and registers a temperature of 14·5° C. (58° F.) On issuing from the rock, the water is clear and limpid; but on contact with the air it soon becomes cloudy, and an ochreous deposit is formed. Although it has a very distinct styptic, inky flavour, it is not unpleasant to the taste, and invalids drink it with pleasure.

CHEMICAL ANALYSIS (SOURCE DOMINIQUE)

WATER 1000 grammes
(1 litre).

Grammes.

Sulphuric acid . .		Sulphuric acid, free . .		1·31	
Arsenic „ . .		Silicate, acid			
Sesquioxide of iron	gram. 1·75,	Arseniate „	Sesquioxide		
Lime and soda . .	thus	Phosphate „	of iron		
Silicic acid . . .	distributed	Sulphate „			
Chlorine		Sulphate of lime		0·44	
Phosphoric acid .		Chloride of sodium . . .			
Organic matter .		Organic matter			

1·75

The *Source Saint-Louis*, which has been more recently discovered, is of the same essential chemical composition, but in a less degree, since the water has only gram. 0·4647 per litre, according to the analysis of MM. O. Henry and Lavigne.

The ferruginous, alkaline, and arsenical water of Vals is aperitive, restorative, tonic, sedative, and also febrifuge, and antiperiodic.

From what we have learnt of the composition and physiological effects of the Vals waters it is easy to understand to what classes of disease they are most applicable. These are dyspepsia, gastralgia, chronic gastritis, gastro-enteritis, and stomach cramp; liver complaints, such as hepatic congestion, jaundice, hepatitis, hypertrophy of the liver, hepatic tumours, biliary calculi, cirrhosis, and fatty degeneration of liver; diabetes and albuminuria; gravel, vesical calculi, chronic cystitis, vesical irritability, prostatic enlargement, &c.; metritis, leucorrhœa, and amenorrhœa; gouty and rheumatic diathesis; chlorosis, anæmia, and neuralgia; enlargement of the spleen and intermittent fever; and also many manifestations of the scrofulous and tubercular diatheses.

These waters can all be preserved perfectly, and are exported in considerable quantities. The strong alkalinity of some of the springs renders them purely medicinal—and not table—waters. They must not be taken in cases of organic disease or general debility without medical advice.

A stay at Vals is very agreeable for patients, as they find there all the comforts which can be desired; and the locality abounds with charming walks and drives.

NEYRAC

(Ardèche)

CALCAREO-BICARBONATED WARM SPRINGS

NEYRAC is a small town built in a picturesque situation on the right bank of the river *Ardèche*, about a mile and a half from Vals, and eight from *Aubenas*. There are seven calcareo-bicarbonated springs, of which only one is used. This is known as the *Source les Bains*, and yields more than 176,400 gallons per day. The water, which has a temperature of 27° C. (81° F.) , supplies a small Establishment containing public and private baths and douches.

The waters are aperitive, tonic, solvent, and detersive. At the time of the Crusades they had the reputation of curing leprosy. At the present day they are strongly recommended for the treatment of certain skin diseases of a parasitical type, such as the different varieties of tinea; also scrofulous swellings, chlorosis, leucorrhœa, and some forms of gastralgia.

This Station is little frequented except by the inhabitants of the Department.

BAGNOLS

(*Lozère*)

SODIO-SULPHURETTED WARM SPRINGS

BAGNOLS is a village of 400 inhabitants, situated at twelve miles from *Mende*, built at a height of more than 2490 feet above the level of the sea, on the slope of a mountain, at the foot of which runs the river *Lot*. The ground is rocky, and abounds in a species of reddish slate, in which are embedded good-sized blocks of quartz.

The waters of Bagnols are sodio-sulphuretted, in-sipid, clear, and colourless, with a slightly styptic taste, and are oily to the touch. At the source they give off almost continually large emanations of gas, the appearance of which is preceded by a curious subter-ranean noise.

CHEMICAL ANALYSIS

WATER 1000 grammes (1 litre).	Grammes.
Carbonic acid	0·323
Sulphuric acid	0·136
Hydrochloric acid	0·035
Silica	0·077
Oxide of iron	0·001
Lime	0·022
Magnesia	0·023
Soda	0·295
Sulphuretted hydrogen	0·027
Traces of phosphoric acid, organic matter, and nitrogen.—*Rivot.*	0·939

The springs, four in number, supply 57,330 gallons of water in twenty-four hours. The most abundant, which is also the warmest, has a temperature of 43° C. (109° F.) The other three are 35° C. -(95° F.), 31° C. (88° F.), 22° C. (72° F.). This difference of tem-perature admits of the preparation of strong, weak, or medium baths and douches, without alteration of the mineral constituents.

The Establishments are two in number; one, an extensive public Institution, comprising six public baths, each capable of containing thirty persons,— douches, vapour baths, an inhaling-room, and a *buvette;*

the other, a private Establishment, smaller, but more comfortably fitted.

The waters of Bagnols act strongly upon the skin and the mucous membranes, the functions of which they stimulate; they produce good effects in rheumatic complaints, scrofulous diseases, and chronic affections of the chest; they have a beneficial effect upon certain skin diseases and atonic ulcers; and are also recommended in traumatic affections, such as gunshot wounds and contusions. In fractures and dislocations with persistent painful swellings, they have proved very beneficial. They resemble in their therapeutical action the sulphuretted waters of the Pyrenees.

CRANSAC

(*Aveyron*)

CALCAREO-SULPHATO-FERRUGINOUS SPRINGS

CRANSAC is a village situated about twenty-two miles from *Villefranche*, and thirty from *Rodez*, in a pretty valley, containing several factories. Near the village is an active volcano, called *le Montet*, which has been in eruption for centuries; steam and acid fumes escape from wide cracks in its sides.

The springs are five in number, but only two are utilised medicinally; the *Source Haute-Richard*, and *Source Basse-Richard*. They are ferruginous, with strong proportions of sulphate of lime and magnesia,

and give 110 gallons of water per day at a temperature of 10° C. (50° F.)

CHEMICAL ANALYSIS

WATER 1000 grammes (1 litre).	SOURCE BASSE-RICHARD.
	Grammes.
Sulphate of potash	0·021
„ soda	0·011
„ lime	2·413
„ magnesia	2·291
„ alumina ⎫	
„ peroxide of iron ⎬ . . .	2·079
„. manganese ⎭	
Chloride of ammonium	0·012
Iodide of ammonium	0·009
Silicic acid . . : . .	0·005
Sulphide of arsenic . : . .	traces
Blondeau.	6·841

The waters of Cransac are clear, without colour or
smell, with an acid and styptic taste. They are free
from gas, and leave an ochreous deposit. The water
of the *Source Basse* is stimulating, diuretic, and laxative. Taken in doses of four or five tumblerfuls, it
has a strong purgative effect. The *Source Haute*, on
the contrary, is constipating, but is more tonic and
restorative.

The waters of the *Source Basse* are recommended
for obstructions of the liver, jaundice, gastric derangement, dyspepsia, chronic gastralgia, constipation, and
congestion of the liver and spleen. It has been claimed
for this spring that it is efficacious against " obstinate
intermittent fevers, improperly treated at first, or

which have not yielded to quinine; as well as against those chronic enlargements of the abdominal viscera which follow protracted fevers." Dr Bras, physician in chief of the Hospital of Villefranche, supports this assertion by his observations upon soldiers returned from Africa, who were cured by these waters of enormous enlargements of the spleen and liver.

At Cransac, crevices in the mountain side are utilised for natural *étuves*, or hot-air baths,—the air being very warm, and loaded with sulphurous vapour. Torpid rheumatism is usually favourably modified by these exhalations, which have this advantage, that they contain no moisture.

CORSICA

This island, from its advantageous geographical position, its climate, and the wonderful fertility of its soil, is a spot specially favoured by nature.

The mineral springs, which are numerous, are for the most part sulphureous. Of this class are Pietrapola, Puzzichello, and Guagno. Those of Pietrapola and Guagno are hot. There are also several cold ferruginous springs, that of Orezza being of such importance as to eclipse all the others.

PIETRAPOLA

(*Corsica*)

WARM SODIO-SULPHURETTED SPRINGS

The valley of Pietrapola, in the canton of Prunelli, is enclosed in the centre of mountains, the aspect of which is most picturesque. The springs, which issue from a plateau in the centre of the valley, are ten in number, and are known as "the Fiumorbo Springs." They are all sulphureous, and have a temperature varying from 32° to 58° C. (90° to 137° F.)

The water is clear, with a distinctly sulphureous taste and smell. It contains gram. 0·025 of sulphuret of sodium to the litre, together with some alkaline salts and chlorides, and is also rich in barégine.

The Establishment comprises three public baths or *piscines*, each capable of containing from thirty-five to forty persons, twelve private bath-rooms furnished with spacious baths, douches, and a large cooling chamber. Close by, is the hotel where patients usually reside. The Pietrapola waters are indicated in cases where there is undue nervous irritability. In certain inter-mittent (not periodical) neuralgias, their use diminishes the frequency of the paroxysms, reducing their acute-ness,—and very often succeeds in affecting a complete cure. Hysteria, chorea, spasms, and certain neuralgic affections of the neck of the womb, usually yield to the action of the waters, the baths being taken at a rather low temperature; and, in addition to nervous diseases, the waters have been used successfully in other cases, especially those of skin disease, contractions of tendons, necrosis, caries, and the results of syphilis.

PUZZICHELLO

(*Corsica*)

COLD SULPHUREOUS SPRINGS

THESE springs are situated near *Casaghianda*, not far from the road which runs along the eastern shore of Corsica. Only two are of importance, which are close together. They are both cold, their temperature being 14° C. (57·2° F.), and have a styptic and nauseous taste. The water contains 3 per cent. of free carbonic acid gas, and some alkaline carbonates, together with a peculiar bituminous substance.

The Thermal Establishment comprises one large public *piscine* and seventeen private baths, an ascending douche, two *buvettes*, and a room appropriated to the use of the slimy deposits from the springs.

Taken in doses of several glasses, this water acts as a slight purgative for the first few days; it also often causes congestion of the venous plexus of the rectum, and even produces a hæmorrhoidal discharge. The waters are almost always administered both internally and externally. The action of the baths is tonic, penetrating, and chiefly affects the skin, which it irritates, reddens, and sometimes covers with true exanthemata. The majority of cutaneous affections are treated successfully at Puzzichello, especially those accompanied by atonic and serpiginous ulcerations. In these cases the action of the bath is powerfully seconded by the local application of the slime, either in its natural state or combined with some fatty substance. The denuded surfaces are covered with it, and very shortly they are stimulated, detergent action sets in, and cicatrisation takes place.

The ruins of *Aleria* are only a short distance from Puzzichello, and a pleasant excursion may be made to visit this ancient city,—which, founded by Sylla, was successively plundered by the different peoples who from time to time conquered without subjugating this island.

GUAGNO

(*Corsica*)

WARM SULPHUREOUS SPRINGS

THE springs of Guagno are situated about forty miles from Ajaccio, in a valley which is traversed by the *Grosso*, one of the principal tributaries of the *Liamone*. The road, for some parts of the way, runs along the sea shore, and is wide and well kept, although there are some very steep places where it crosses several chains of mountains.

The Thermal Establishment, which includes the Military Hospital, can accommodate more than 300 patients. There are two springs, which are connected at the point where they issue from the ground. Their temperature is 52° C. (126° F.) One portion of the water passes directly to the douches; the other is carried into two large tanks, whence (after being cooled to the necessary temperature) it is distributed to the *piscines* and private baths. In the centre of the main building is the *buvette*.

The water gives off a slight smell of sulphuretted hydrogen, and has a faint and sickly taste. According to M. Poggiale, it contains gram. 0·024 of sulphide of sodium to the litre; but this analysis would require to be confirmed.

The waters are efficacious in certain cutaneous affections, particularly eczema in its different forms. They have also been of benefit in cases of simple or complicated articular rheumatism, and sciatic neuralgia. Certain forms of paralysis, gun-shot wounds, and the results of syphilis, have often been advantageously treated at Guagno. The waters are thus somewhat analogous to those of Barèges.

Guagno is surrounded on all sides by mountains, covered with immense forests, the gloomy and mysterious aspect of which—added to the fact that one is, so to say, in the centre of the country made classical by the exploits of the early bandits—cannot fail to

impress the visitor. Theodore, one of the most celebrated of these heroes, was born here, and his feats
are still related by the inhabitants.

OREZZA

(*Corsica*)

COLD FERRUGINOUS SPRINGS

THESE springs rise in the canton of *Piedicroce*, at
the bottom of a charming valley, on the right bank
of the river *Fiumalto*. The water, which bubbles
and foams up, is cold, its temperature being only
14° C. (57·2° F.)

M. Poggiale, in his analysis, states that it contains
gram. 0·128 per litre of carbonate of iron, and gram.
1·248 of carbonic acid gas.

CHEMICAL ANALYSIS

WATER 1000 grammes (1 litre).	
	Litre.
Carbonic acid, free	1·248
Atmospheric air	0·011
	Grammes.
Carbonate of lime	0·602
„ magnesia	0·074
„ and oxide of iron . . .	0·128
Sulphate of lime	0·021
Chloride of potassium }	0·014
„ sodium }	
Silicic acid	0·004
Alumina	0·006
Traces of carbonates of lithia, of oxide of man-ganese and cobalt, and of arsenic acid.—*Poggiale.*	0·849

The chemical composition of Orezza water is obviously very remarkable, and the medicinal properties are not less so. Very good results are obtained by its use in cases of chloro-anæmia, leucorrhœa, passive hæmorrhages, and chronic diarrhœa, especially when arising from atony of the mucous membrane.

Up to the present this water has not always been successfully transported, but it has often become thick and troubled. It is to be hoped, however, that as its consumption increases yearly, and its value is more widely known, some means may be discovered of preserving it in a better condition.

EASTERN DIVISION

In this part of the country the mineral springs rise chiefly in mountain districts; but the mountains are not extinct volcanoes like those of Auvergne, they are richly wooded and fertile. The most important springs are to be found in the Vosges and Dauphiné.

BOURBONNE-LES-BAINS

(*Haute-Marne*)

SODIO-CHLORINATED WARM SPRINGS

BOURBONNE-LES-BAINS is a small town, pleasantly situated on a little hill, at a height of about 900 feet above the level of the sea. It stands in a pretty valley, watered by two streams called the *Borne* and the *Apauce*, and commands a view of the Vosges mountains in the distance.

Until lately there were only three springs—the *Source de la Place*, temperature 58·75° C. (138° F.), which supplied the *buvette*; the *Source Puisard*, tem-

14

perature 57·75° C. (136° F.), supplying the Thermal Establishment; and the *Source des Bains-Militaires*, temperature 50° C. (122° F.), supplying, as its name indicates, the Military Hospital. But since the year 1857 new borings have been made, and fresh springs brought into use. They all contain the same mineral components, and differ only in the quantities of these substances and in temperature, which varies from 60° to 65° C. (140° to 149° F.)

The waters of Bourbonne-les-Bains are sodio-chlorinated, and are given off at the rate of 110,250 gallons per day. They are clear, colourless, and full of gas bubbles, which rise to the surface and burst, so that they have almost the appearance of being always boiling. They are very salt, and their flavour is bitter, not at all nauseating, and followed by a faint after-taste, somewhat resembling that of weak veal broth. They at first feel smooth and pleasant to the skin, but soon produce a slight sensation of dryness and stiffness.

Bourbonne contains both a Civil and a Military government Hospital for the use of the waters. The latter can accommodate 100 officers and 300 soldiers.

The Thermal Establishment has been reconstructed on a new plan; it now consists of two completely separate buildings, for first and second class baths respectively.

The first-class Establishment comprises eighty-two bath-rooms (there being no public bath), and fifty-seven douches, of which two are in the baths, two circular, and two ascending. The second class establishment includes six enormous *piscines* or public baths, and has no private bath-rooms, except a few for such patients as cannot bathe with others on account of the nature of their complaints. There are also fourteen douches.

The waters are used for drinking, and for baths, douches, injections, fomentations, vapour-baths, &c.

Several analyses of their composition have been made at different times; the most recent, by M. Pressoir, chief chemist to the military hospital, is as follows :

CHEMICAL ANALYSIS

WATER 10,000 grammes (10 litres).	Grammes.
Chloride of sodium	58·00
„ magnesium	4·00
Carbonate of lime	1·00
Sulphate of lime	8·80
„ potash	1·30
Bromide of sodium	0·65
Silicate of soda	1·20
Alumina	1·30
Peroxide of iron	0·03
Manganous manganic oxide	0·02
Traces of iodine and arsenic.—*Pressoir.*	76·30

When drunk very hot, the thermal water stimulates the functions of the stomach and intestines, and increases their secretions; it has a very distinct action upon the cutaneous system, and also often produces an abundant flow of urine. At a moderate

temperature, of from 15° to 20° C. (59° to 68° F.),
it has generally a slightly laxative action, which is of
great advantage in cases of obstruction of the abdominal viscera, and affections of the nervous centres.
The duration of the bath varies from forty-five
minutes to one hour, and its temperature from 32°
to 36° C. (90° to 97° F.) Its immediate effect is to
stimulate the skin and increase its vitality (even to
the point of producing congestion), and to accelerate
the superficial circulation. The douche is one of the
most active agents in the cure ; besides its stimulating
and resolvent action, it produces a beneficial effect
by its vigorous percussion on the tissues, which, in
fact, is real *massage*. Fomentations are employed
when the gravity of the disease does not permit the
use of the douche.

For some years electricity has been used at this
Station, in connection with the thermo-mineral treatment, and excellent results have been obtained,
especially in cases of rheumatism and paralysis.

The waters of Bourbonne-les-Bains may be recommended in all cases of scrofulous diathesis, from a
simple predisposition to this malady up to its most
serious forms—osteitis, caries, knee and hip-joint
diseases ; also for arthritic and syphilitic constitutions,
and the cutaneous affections (though for the last
sulphurous waters are preferable,) connected with
these three diatheses ; in slow convalescence after
dangerous fevers ; for some forms of gastro-intestinal
dyspepsia, and certain varieties of paralysis and neu-

ralgia ; and, finally, in cases of fractures, dislocations, sprains, strains, hydrarthrus, anchylosis, contractions and atrophies, as well as gun-shot and other wounds.

They are distinctly to be avoided in all acute complaints, and in affections of the chest, heart, and large blood-vessels.

Life at Bourbonne is not at all so serious a matter as might be supposed from the class of diseases treated there. The Establishment possesses a casino, where two balls are given every week, and there is dancing also on the other evenings. Visitors who wish to explore the neighbourhood ought to see the ruins of *Coiffy-le-Haut,* the abbeys of *Morimond, Vaux-la-Douce,* and *Flabemont,* and the famous oak called *Chêne des Partisans,* in the Forest of *Saint-Ouen,* which is four hundred years old, eighty-two feet high, and forty-six feet in circumference.

PLOMBIÈRES

(*Vosges*)

SODIO-SULPHATED WARM SPRINGS

PLOMBIÈRES is a small town of 1500 inhabitants, situated at a height of 1760 feet above the level of the sea, in a deep valley on the southern border of the Department of the Vosges. It lies from west to east, along both banks of the Augronne, a real torrent, partly hidden from sight by an arch, at the feet of two high mountains which hem it in between

their approaching slopes.   When one approaches
Plombières by the road from Épinal, the first thing
seen is the smoke rising from the still invisible town,
so that the traveller feels as if he were about to
descend into the enormous crater of some volcano;
but soon, as he advances down the steep slope which
leads him to his destination, he distinguishes roofs of
houses, and charming cottages, picturesquely dropped
down among the meadows that adorn the steep slopes
of mountains, the summits of which are crowned with
forests.   Nothing is more singular or more strangely
beautiful than the view of Plombières, seen thus from
the crest of the hill which has to be descended on the way
from Épinal ; then, when the visitor has at last reached
the bottom of the valley, and turns sharply to the
right, he has before him the chief part of the town,
that is to say, a single, wide, well-paved street, with
foot-paths, along which rise houses, generally of an
elegant style, and which, in the season, are transformed
into furnished lodging-houses for the accommodation
of visitors.

There is perhaps no Thermal Station which possesses
so many hydropathic establishments as Plombières.
There are six, known respectively as *Bain Impérial,
Bain Tempéré, Bains des Capucins, Bains Romains,
Bain des Dames,* and *Thermes Napoléon.*   All these
include several public baths, numerous private baths
and douches, as well as hot-air baths.   The most
important is incontestably the *Bain Napoléon ;* and,
without entering into details, it may be said that since

it has been constructed, the town of Plombières has possessed the most complete and admirably organised bathing arrangements that can be desired. The *Bain Romain*, formerly called *Bain des Pauvres*, however, carries off the palm of elegance. Besides these six establishments, there are four *buvettes*, and a Hospital which receives, on an average, every year, more than 300 patients, civil and military.

The waters of Plombières belong to the sodio-sulphated class. They are clear and colourless, without smell, almost tasteless, and oily to the touch. Their temperature varies between 15° to 71° C. (59° to 160° F.), and their chemical composition is almost identical, so that it will be sufficient to give here the analysis of the *Source des Dames*.

CHEMICAL ANALYSIS

WATER 1000 grammes (1 litre).	SOURCE DES DAMES.
	Cubic cent.
Oxygen	1·77
Nitrogen	9·62
	Grammes.
Carbonic acid, free . . .	0·01267
Silicic acid	0·02731
Sulphate of soda . . .	0·09274
Bicarbonate of soda . . .	0·01125
„ potash . . .	0·00133
„ lime . . .	0·03868
„ magnesia . . .	0·00670
Chloride of sodium . . .	0·00927
Silicate of soda	0·05788
Traces of sulphate of ammonia, arseniate of soda, silicates of lithia and alumina, fluoride of calcium, oxide of iron and manganese, and nitrogenous organic matter.—*Jutier and Lefort.*	0·25781

As appears from this analysis, these waters are faintly mineralised, yet their therapeutical properties are very valuable. Their successful use is not only due to the richness of their thermality, but also to the manner in which they are administered.

The treatment at Plombières is both external and internal. All the springs are used in the former, but only those *des Dames* and *du Crucifix* for the latter. In whatever manner administered, they are slightly stimulating, diuretic, aperient, tonic, and restorative.

There is also a ferruginous spring called *Bourdeille*, the water of which is drunk only.

If there is any one among mineral waters which acts directly upon the intestines, it is that of Plombières. Thus, gastro-intestinal dyspepsia often yields, as if by magic, to the well directed use of this water; but its speciality is the cure of inveterate diarrhœa. Improvement usually begins to be noticeable only after the fifth or sixth bath. This action of the waters explains why many women—with whom constipation has become, in a manner, a normal condition—are obliged, during their course at Plombières, to make use of the ascending douche; and probably accounts for the fact that these douches were there first introduced.

These waters are highly recommended for irregularities of menstruation characterised by a want of vitality of the uterus, in leucorrhœa from atony, and sterility resulting from the same cause. They are also prescribed with advantage for different nervous affections, such as headache, sciatic or facial neuralgia, chorea, and certain forms of paraplegia resulting from irritation of the spinal cord.

Some forms of cutaneous diseases of a herpetic nature, which are still in a subacute stage, can also be beneficially treated with these waters.

The bathing season opens officially on May 15th, and ends with the month of September; but the particular period during those four months at which the waters should be used is not a matter of in-difference. Rheumatic, gouty, or neuralgic patients, ought to be treated in the early part of the season, in order that the chills of autumn may not strike upon them while under the influence of the thermal stimulation. Persons suffering from nervous and stomach complaints will not be the worse for undergoing their course of treatment when the season is more advanced; while with regard to uterine and cutaneous diseases, the question is of no importance.

Plombières is a much frequented watering-place, offering many attractions to visitors. First there is the Casino, the arrangements of which leave nothing to be desired; then the promenades *des Dames, de la Feuillée Dorothée*, and *Saint Loup;* and finally the numerous and delightful excursions in the neighbourhood. Among the best may be named those to the *Fontaine Stanislas*, the *Val d'Ajol*, the valley of *Semouze, Remiremont* and its old ruined abbey, &c.

LUXEUIL

(*Haute-Saône*)

SODIO-CHLORINATED SPRINGS, AND FERRUGINOUS WARM SPRINGS.

LUXEUIL is a small town, with a population of 4000, situated at the foot of the Vosges mountains, in a pleasant plain watered by two streams, the *Breuchin* and the *Lantène*. To judge by the ruins of its ancient baths, Luxeuil must at one time have equalled and even surpassed Plombières in magnificence. There is also a curious analogy in the fortunes of these two baths. Both were destroyed by Attila, and centuries later both were restored simultaneously, though in different styles. While at Plombières everything is sumptuous, the arrangements at Luxeuil have a certain air of almost coquettish elegance, in harmony with the tastes of the ladies who form the majority of the visitors. Mirrors and sculptures are seen on every side, and every bath-room is fitted up like a boudoir.

Since the latest explorations, the mineral springs of Luxeuil number eighteen, but all are not yet utilised. The following are the names and temperatures of the most important :—*Grand-Bain*, 56° C. (132° F.) ; *Source des Cuvettes*, 44° C. (111° F.) ; *Bain Gradué*, from 35° to 38° C. (100° F.) ; *Bain des Fleurs*, 38° C. (100° F.); *Eau savonneuse*, 30° C. (86° F.); *Bain*

des Dames, 47° C. (116 F.); *Bain des Bénédictins,* 37°
C. (98° F.); *Source d'Hygie,* 29° C. (84° F.); *Fontaine
des Yeux,* 29° C. (84° F.). These various springs have
been re-installed in the old Establishment, which has
been enlarged and thoroughly renovated. It comprises now seven divisions, named after the springs
which supply them.

This almost too elegant Establishment is situated at
the northern end of the town, in the midst of a large
garden, and is indisputably one of the finest of its kind
in France. It possesses three great *piscines,* or public
baths, in which 130 persons can bathe at the same
time in water incessantly renewed; seventy private
bath-rooms (of which ten are double), and forty bath-
rooms where the douche can be received in the bath;
all perfectly fitted up, and furnished with taps of hot
and cold water. There is, besides, a complete assort-
ment of douches, varied both in strength and temperature, the shower douche, Scotch douche, ascending
douche, apparatus for vaginal and uterine irrigation,
besides rooms for general and local vapour-baths,
inhalations, &c.

The waters (except that of the ferruginous springs,
of which we shall speak separately), are clear, inodorous, and somewhat oily to the touch; they have hardly
any perceptible taste. The most recent analysis shows
that the spring of the *Bain des Dames* (which is the
most strongly mineralised) has almost the same composition as the waters of Plombières, except that it
contains more manganese.

CHEMICAL ANALYSIS

WATER 1000 grammes (1 litre).				SOURCE DU BAIN DES DAMES.
				Cubic cent.
Oxygen	.	.	.	2·36
Nitrogen	.	.	.	7·54
Carbonic acid, free	.	.	.	20·84
				Grammes.
Sesquicarbonate of potash	.	.	.	0·04350
Chloride of potassium	.	.	.	0·02589
Sulphate of soda	.	.	.	0·13716
Chloride of sodium	.	.	.	0·72333
Carbonate of lime	.	.	.	0·03859
„ magnesia	.	.	.	0·00215
Fluoride of calcium	.	.	.	0·01385
Silicic acid	.	.	.	0·09810
Organic matter .	.	.	.	0·02589
Traces of iodine and manganese.—*Lecomte.*				1·10846

The ferruginous waters come from three springs, of which the following are the names, temperatures, and the quantitative proportions of sesquioxide of iron contained in one litre (1000 grammes). The *Temple*, 19·6° C., gram. 0·02500; *Puits Romain*, 27·9° C., gram. 0·000939; *Pré-Martin*, 19° C., gram. 0·02500.

In consequence of these diversities of mineralisation, the therapeutical resources of Luxeuil are extremely varied. The alkaline waters should be prescribed for dyspepsia and gastralgia; for rheumatism with predominance of the nervous symptoms; also for sciatica and hysteria. The affections which generally attack lymphatic constitutions, where a slow circulation, loss of tone, colourless skin and mucous membranes, point clearly to chlorosis and anæmia, should be treated with the ferruginous waters; among them it is enough to mention spermatorrhœa, blennorrhœa; leucorrhœa, amenorrhœa, and excessive menstrual discharge resulting from want of contractile power of the uterus or general debility; also congestions or displacements of the womb, &c.

Finally, like many other Thermal Stations, Luxeuil claims for its waters the credit of curing sterility. The different authors who have written upon these springs have noted several cases of ladies who have had children after a season passed at Luxeuil, though up to that time they had been disappointed in their natural hopes.

Reading and conversation saloons, ball and concert-rooms, are all provided for the amusement of visitors to Luxeuil, and charming walks can be had in the neighbourhood, which is very picturesque. The old town will be explored with interest, and the abbey, of which fine ruins still remain, was one of the most celebrated of the Middle Ages, and is rich in historical associations.

BUSSANG

(*Vosges*)

FERRUGINOUS COLD SPRINGS

BUSSANG is a little town of 2500 inhabitants, situated on the boundaries of the departments of the Vosges and the Bas-Rhin, almost at the source of the Moselle, at the foot of one of the highest mountains of the Vosges range, called the *Penhaut.*

Its mineral waters are obtained from springs situated a little over a mile outside the town, in a glen of the valley of the Moselle, at the foot of the *Charet* Mountain. The waters are ferruginous, cold, colourless, and generally quite clear. After being at rest for some time, they form at the bottom of the vessel a brown deposit of oxide of iron, which is found in abundance on the inside of the reservoirs, and on the pavement round the springs. In the reservoirs the water does not appear transparent, for a thin coating forms on the surface, which reflects the light in prismatic colours; it bubbles, as if just going to boil, in consequence of the large quantities of carbonic acid that are constantly escaping. It has a fresh, sharp taste, which is at first agreeable to the palate, but is succeeded by a metallic flavour due to the iron it contains.

CHEMICAL ANALYSIS

WATER 1000 grammes (1 litre).	SOURCE D'EN BAS.
Carbonic acid, free . . . Litre	0·41
	Grammes.
Carbonate of soda	0·789
„ lime	0·340
„ magnesia	0·150
„ iron	0·017
Crenate of iron, manganese, and traces of chloride of sodium	0·078
Sulphates of soda and lime . .	0·110
Silicate of soda	0·002
Traces of carbonate of strontia and crenate of soda.—*O. Henry.* .	1·486

Since the above analysis was made, the presence of arsenic in the waters of Bussang has been demonstrated; but it exists there in very small quantities, only about three milligrammes to the litre (1000 grammes).

These waters are aperitive, tonic, alterative, and remarkably facilitate digestion. They are successfully used in cases of dyspepsia, gastralgia, chlorosis, hysteria, anæmia, cardiac and vascular murmurs, &c.

There is no Thermal Establishment, for as the waters can be sent to any distance without undergoing any deterioration, they are chiefly drunk at home, and enjoy a considerable reputation.

CONTREXÉVILLE

(*Vosges*)

CALCAREO-SULPHATED AND BICARBONATED COLD SPRINGS

CONTREXÉVILLE is a village of 700 inhabitants, situated in a funnel-shaped valley which runs from south to north, and is watered by the little river *Vaix*. It owes its reputation to its mineral waters, which were discovered in 1759 by Dr Bagard.

The springs are four in number, and are called respectively the *Source du Pavillon, du Prince* (or *des Bains*), *du Quai*, and *la Souveraine*. The first is used exclusively for drinking; it yields no less than 13,230 gallons per day; its temperature is 11·5° C. (52·25° F.),

and its density is 1·029, a little higher than that of
distilled water. The two next are reserved for the
supply of the baths and douches.   All these three have
been known as long as Contrexéville has been recognised as a watering-place, and belong to the class of
calcareo-sulphated waters.   The *Souveraine* spring
differs considerably from the others in chemical
composition.

CHEMICAL ANALYSIS

WATER 1000 grammes (1 litre).					SOURCE DU PAVILLON.
					Grammes.
Carbonic acid, free		.	.	.	0·080
Bicarbonates { of lime	.	.	.	.	0·402
of magnesia	.	.	.	.	0·035
of iron	.	.	.	.	0·007
of lithia	.	.	.	.	0·004
Sulphates { of lime	.	.	.	.	1·165
of magnesia	.	.	.	.	0·236
of soda	.	.	.	.	0·030
Silica .	.	.	.	.	0·015
Chlorides { potassium	.	.	.	.	0·006
Sodium	.	.	.	.	0·004
Traces of fluoride of calcium and arsenic.— *Debray.*					2·384

The Contrexéville waters are cold, clear, and colourless ; they deposit in the pipes and reservoirs an oily
and ochreous sediment, and on contact with the air
their surface becomes covered with an iridescent
coating. Their taste is fresh, acid, somewhat inky,
and leaves behind a styptic after-taste. When taken
internally, they are rapidly absorbed. Their presence
in the vascular system is manifested by a quickened

pulse, greater frequency of respiration, and increased activity of all the secretions, especially those of the bladder and intestines. They are in a high degree laxative; and a few hours are sufficient for the quantity drunk to be eliminated by the kidneys and expelled by the intestines.

The analysis of the *Pavillon* spring shows that it gives a sulphated and bicarbonated water, containing lime, magnesia, and lithia—to which iron and arsenic add their restorative qualities.

The springs *du Prince* and *du Quai* have the same temperature as the *Pavillon*; the former gives $2\frac{1}{2}$ and the latter over 14 gallons per minute. Their physical properties are the same as those of the *Pavillon*; but their chemical constituents differ; the *Source du Prince* containing more iron and arsenic, and the *Source du Quai* more magnesia than the first-named spring.

The *Souveraine* spring was discovered about twenty years ago; its temperature is only 10° C. (50° F.); it contains no iron, but on the other hand is richer in magnesia; and it is less diuretic, but more laxative.

The Contrexéville waters have the property of stimulating all the secretions, but especially that of the kidneys. They differ from those of Vichy in being suited to all forms of gravel, instead of—like the latter—to only one, since they act rather by a sort of repeated irrigation than by chemical combinations. They produce considerable diuresis, and cause the excretion of a quantity of liquid larger than that of the water drunk; they modify the mucous membrane

15

of the urinary passages in a manner which accounts for their remarkable success in cases of vesical catarrh. They cause the painless expulsion of gravel from the kidneys, and the excretion of any excessive quantity of uric acid in the system. Lastly, they reveal the presence of calculi hitherto undetected, whose size does not allow of their passage through the natural channels.

The successes obtained at Contrexéville in the cure of gravel are universally recognised, and the brilliant results which occasioned the saying that " Contrexéville is the antidote of gravel," would be still more marked if all persons liable to this complaint would continue to submit to the hygienic directions given them at their departure. Unfortunately, they often return to too rich a diet, or too sedentary a life ; and then the same causes reproduce the same effects, and they are forced to go back, in order to seek at the *Pavillon* spring immunity for future imprudences. These remarks apply especially to the uric acid (or red) gravel.

In phosphatic (or white) gravel, no treatment has given better results than those which have been obtained at Contrexéville. In this complaint, the urine becomes either strongly alkaline, or at least less acid than in a normal state ; and—what will at first seem extraordinary—under the influence of this faintly alkaline water the urine recovers its acidity. This results from the fact that, as the inflammation of the urinary passages abates, the secretion gradually returns to its normal condition. When phosphatic calculi are present, they are often expelled in the form of soft paste, for they are far from having the hardness of uric acid calculi, but are reproduced much more quickly.

Gout, and above all *atonic gout*, is one of the affections which fall within the province of the Contrexéville waters. If any one doubted the connection of gout and gravel, he would be soon enlightened by a season at this watering-place, which is *par excellence* the station for gouty patients suffering from gravel, and gravel patients suffering from gout.

After gout and gravel, the diseases which are most frequently under treatment at this station are vesical catarrh; atony, inertia, and paralysis of the bladder; retention and incontinence of urine. "The Contrexéville waters," said Dr Civiale, "are specially successful in restoring the contractility of the bladder, which is almost always weakened by chronic catarrh of this organ. Several of my patients to whom I had prescribed these waters were so much benefited by their use that they returned to them of their own accord." To this series of diseases may be added prostatic enlargement and chronic cystitis, for both of which these waters are also beneficial. Great advantage has been derived from a course of Contrexéville waters, by persons who have undergone lithotrity, and those who have been cured of long-standing strictures of the urethra.

Their action on the intestines is laxative without being debilitating. From two to six stools is the experience of almost all patients under treatment.

These evacuations do not reduce the quantity of urine, which sometimes appears greater than the quantity of water that has been drunk. It would appear probable that a large quantity of water introduced into the stomach would be liable to fatigue it, but on the contrary, in most cases the appetite is notably increased, and digestion becomes more easy and rapid.

Up to the present time, baths and douches have only played a secondary part in the Contrexéville treatment, probably in consequence of the want of suitable accommodation. Now that they have been reorganised on a very large scale, with all modern improvements, they constitute—on the contrary—an important feature of the treatment. Forty-eight bath-rooms, five large douches, three hip-baths with running water, one vaginal, and two ascending douches, provide amply for all requirements at this station. While the baths cause a general relaxation of the system, the douche directed upon the loins arouses a slight shaking in the region of the kidneys, which encourages the passage of the calculi into the bladder, and consequently their expulsion.

These waters keep perfectly, and are exported in considerable quantities.

Contrexéville is far from being a dull place of residence, but it must be remembered that the patients who frequent it—and many of whom return every season— are usually no longer young, and are scarcely disposed for noisy pleasures or fatiguing amusements. Contrexéville may be said to be an annual rendezvous of the cleverest men, the wittiest talkers, and the most

noted *gourmets* that one common malady can assemble.
Patients are constantly meeting each other, and spend
the greater part of the day in company; so that all the
agreeable social qualities which they possess are needed
to prevent mutual weariness and boredom.   Reading,
walking, games, and concerts, are the only amusements
to be had.   Moreover, the patients get well.   Can they
ask for more ?

VITTEL

(*Vosges*)

MIXED SULPHATED SPRINGS, AND FERRUGINOUS COLD SPRINGS.

VITTEL is a village standing in a valley watered by
the river *Vair*, at the foot of a wooded mountain,
about three miles from Contrexéville, and at a height
of 1100 feet above the level of the sea.   It owes its
celebrity entirely to its mineral waters, which are quite
a modern discovery, having only been known for about
twenty years.

The mineral springs rise at some distance from the
village, in the midst of a park, in a very healthy
situation, whence the view of the most picturesque
portion of the Vosges chain of mountains is extremely
fine.

The Thermal Establishment comprises, besides
complete arrangements for baths and varied douches,

a large and handsome covered gallery, 144 feet long, where patients can take exercise when the weather does not allow of walks in the park; a public sitting room, where newspapers and games are provided, and a reading-room.

It is supplied by four springs of slightly different composition, called respectively the *Grande Source, Source Marie, Source Salée,* and *Source des Demoiselles.*

CHEMICAL ANALYSIS

WATER 1000 grammes (1 litre).	GRANDE SOURCE.
Carbonic acid, free . . $\frac{1}{10}$ of the volume	Grammes.
Bicarbonate of lime	0·185
„ magnesia ⎱ . . .	0·079
„ soda ⎰	
„ and oxide of iron . . .	0·010
Sulphate of lime	0·440
„ magnesia	0·432
„ soda	0·326
Chloride of sodium ⎱ . . .	0·220
„ magnesium ⎰	
Silica, alumina ⎫	
Phosphate of lime ⎪	
Potassic and ammoniacal salts ⎬ . ′ .	0·047
Iodides and arsenical principles, traces ⎪	
Organic matter ⎭	
Traces of bicarbonate of manganese and sulphate of strontia.—*O. Henry.*	1·739

The first of these, the *Grande Source,* yields at present more than 264,480 gallons per day. The water, whose temperature does not exceed 11° C. (52° F.), is fresh and clear, of a slightly inky smell, and a sharp, somewhat astringent taste. Bubbles of carbonic acid escape from it, and after a time its surface is covered

with a thin prismatic coating. An ochreous deposit
is left upon the interior of the reservoir containing it.

The *Source Marie* is received in a hexagonal basin,
protected by a pavilion of the same shape. This
building has a gallery, commanding a very pretty
view. The spring gives 286,520 gallons of water per
day, at a temperature of 11·50° C. (52·5° F.) The
analysis of this spring is similar to that of the preceding, but it is richer in magnesia; it has a dull and
bitter taste, and its laxative action is more pronounced
than that of the water of the *Grande Source.*

The *Source Salée* yields twenty gallons per minute,
and its mineralisation is superior to the two preceding
ones, and richer in salts of magnesia; its temperature
is 11° C. (52° F.) ; it is specially employed in diseases
of the liver, in gastro-intestinal affections, accompanied
with constipation, and in habitual constipation from
whatever cause ; also in abdominal plethora, accompanied with obesity. It acts as a purgative with some
patients, but is strongly laxative in all cases, and its
action is as manifest on the lower part of the intestines as it is on the commencement of the digestive
apparatus.

The name of the *Source des Demoiselles* was probably
supposed to require greater elegance in the surroundings of this spring than was thought necessary for the
others. It is enclosed in an elegant kiosk, which in
its turn stands in a flower garden. It is an excellent
ferruginous water, containing gram. 0·040 per litre of
bicarbonate and crenate of iron. There are also the

same constant elements as in the *Grande Source* and
the *Source Marie ;* while as to their proportion, the
Source des Demoiselles occupies a middle place between
the other two.

From the differential analyses of these three springs,
it appears that the first is calcareo-sulphated, the
second calcareo-sulphated and magnesian, and the
third bicarbonato-ferruginous. The first resembles in
its composition the *Pavillon* spring at Contrexéville;
it is eminently diuretic, does not in any way fatigue
the stomach, and is very easily digested. The second
is, like the first, diuretic; but it has also a tolerably
strong action upon the intestinal secretions, and produces purging. The third (in consequence of the iron
and traces of iodine which it contains) is tonic and
restorative.

The waters of both the *Grande Source* and the
Source Marie are very efficacious in the treatment of
gravel, gout, chronic cystitis, vesical catarrh, and
strictures of the urethra recently operated on; while
those of the *Source Marie* are especially suitable in
dyspepsia, chronic enteritis, and constipation. The
waters of the *Source des Demoiselles* can be prescribed
successfully in cases of chlorosis, anæmia, amenorrhœa,
leucorrhœa, and in general, in all affections where
iron is likely to be useful.

All these waters are exported, and keep without
deterioration.

Life at Vittel is quiet but not dull, and fêtes and
concerts are given from time to time.

The excursions in the neighbourhood of Vittel are most varied and charming; good horses and excellent carriages are provided by the Establishment.

MARTIGNY-LES-BAINS

(*Vosges*)

CALCAREO-SULPHATED COLD SPRINGS

MARTIGNY is a village of some importance, situated on the road from Contrexéville to Bourbonne-les-Bains, in the middle of a valley watered by the river *Mousson*, and surrounded by hills covered with woods and vineyards.

The three mineral springs which it possesses are distinguished by numbers. They are cold, the water is clear, without smell, and almost without taste, except for a slight astringency.

The use of the waters of Martigny was only authorised in the year 1859, and the Thermal Establishment is consequently of recent construction. It is placed at the end of the village, and stands in a large park.

The springs are calcareo-sulphated, and M. Jacquemin's analyses, made in 1862, show that they have a great analogy to those of Vittel and Contrexéville. That, however, which chiefly distinguishes them from the latter is the greater quantity of lithia, the water of Martigny containing three centigrammes per litre of chloride of lithium, whereas the same quantity of Contrexéville water contains barely four milligrammes of bicarbonate of lithia. All doctors know the experiments with lithia, made by Dr Garrod, who, in his admirable work on 'Gout,' has most distinctly demonstrated the solvent action of this alkaline oxide, upon uric acid, and the urates, chiefly urates of soda, which form the basis of the sediments of red gravel, and of tophi or chalk-stones in gouty persons.

The waters of Martigny would, therefore, be chiefly prescribed for gravel, especially the red, or uric acid, and the different manifestations of the gouty diathesis. They can also be recommended in affections of the urinary passages, such as chronic cystitis, vesical catarrh, vesical atony, hæmaturia, obstinate chronic urethral discharges, diabetes, and albuminuria; enlargements of the liver, spleen, and of the prostate gland; chronic gastritis, dyspepsia, and gastralgia; rheumatic and neuralgic affections, &c.

They are chiefly taken internally, in doses of from two to eight glasses, according to the case, and are sometimes prescribed also in baths and douches. They can be exported without losing any of their essential properties.

BAINS

(*Vosges*)

SODIO-SULPHATED WARM SPRINGS

BAINS is a charming little town with a population of 3000, built 1000 feet above the level of the sea, at the foot of the southern slope of the Vosges mountains, in a valley watered by the *Sémouse,* an affluent of the Saône. It is situated between Contrexéville and Plombières, on the edge of the magnificent forest of Tremonsey.

The numerous springs are all sodio-sulphated; they are enclosed in two Thermal Establishments. The *Bain Neuf* or *Bain de la Promenade,* is a large building somewhat like a hospital. It contains three public baths or *piscines* of an oval shape, arranged in order one after the other, the water in each being of a different temperature. There are also private baths and douches. The *Bain vieux* or *Bain Romain* is a building remarkable for its architecture, which is dignified and simple, and in its general effect reminds the visitor of the ancient Roman Thermæ. On the ground floor there are three elegant public baths, the water of which is constantly renewed. Here, according to the primitive and innocent custom of the Vosges, both sexes bathe together, *en costumes de bain.* The private bath-rooms are on the first floor.

These two Establishments are supplied by thirteen springs, which yield a total of 4,409,200 gallons per day. The principal are the *Grosse Source,* the *Source Savonneuse,* the *Source de la Promenade,* and the *Source de la Vache.* Their temperature varies between 29° and 50° C (84° to 122° F.)

The waters of Bains are clear, without smell, and almost without taste. They are drunk, and used for baths and douches. The two springs which supply the *buvette* or drinking fountain, and which are almost the only ones used for drinking, are the *Source Romaine* and the *Source de la Vache.* Like all waters which are faintly mineralised, those of Bains have no very distinctly specialised therapeutical action. They act chiefly by their thermality. They are very beneficial to persons who have lost tone and strength, and whose digestive powers are impaired; to those affected with erratic rheumatic pains; in false anchylosis of joints and chronic arthritis; and may also be used with good results in some uterine complaints, &c.

SALINS

(*Jura*)

SODIO-CHLORINATED COLD SPRINGS

SALINS is a town containing 7000 inhabitants, situated by the side of a stream called the *Furieuse,* at an elevation of more than 1200 feet above the level of the sea, between the mountains of Belies and St André, each of which is crowned by a Fort.

The Thermal Establishment stands in the centre of the town, at the very foot of the hill on which stands the fort St André. Its arrangements are all that could be desired, and are not surpassed by those of the finest establishments of the kind in France. They include everything requisite for hydro-mineral treatment. There is, first, an enormous *piscine* or public bath, of a circular shape, deep enough (more than four feet) to render swimming possible for those who like it, and with a very high roof which secures a free circulation of air. It contains nearly 20,000 gallons of water, constantly renewed. There are also numerous bath-rooms, several douche rooms, and a complete set of hydro-therapeutical apparatus.

The waters of Salins are sodio-chlorinated, cold. They proceed from several springs, of which only one is used medicinally, while the others are reserved for the manufacture of salt. The medicinal spring is situated in the centre of the Thermal Establishment. It gives daily nearly 4,000,000 gallons of water, which is received in an underground reservoir. It is pumped up by a powerful hydraulic machine, set in motion by the *Furieuse* stream, and distributed to the different parts of the Establishment.

The most recent analysis is that made by M. Reveil, which is as follows :

CHEMICAL ANALYSIS

WATER 1000 grammes (1 litre).	Grammes.
Bromide of potassium	0·03065
Chloride of potassium	0·25662
„ magnesium	0·87012
„ sodium	22·74515
Sulphate of lime	1·41606
„ potash	0·68080
Traces of iodide of sodium and carbonates of lime and magnesia.—*Reveil.*	26·00000

Besides this spring, the concentrated waters (*eaux*

mères), which belong to the salt manufactory, are also utilised. They are brought from the works to a leaden reservoir in the Establishment, from which they are conveyed in buckets of a fixed size to the baths which require strengthening.

The waters of Salins are used both internally and externally. They can be drunk fasting, either before, during, or after the bath. When taken in doses of from one to two glasses, they are digestive, resolutive, tonic, and restorative. In larger doses they become exciting and purgative, and often even produce vomiting, quickening of the pulse, and headache.

The baths constitute the most important part of the treatment. Their duration varies from forty-five to sixty minutes, and their temperature from 24° to 35° C. (75° to 95° F.)

No general rule, however, can be laid down; the temperature and duration of the bath must depend upon the constitution and the particular complaint of the patient. Nevertheless, the temperature of the bath ought to be, as far as possible, moderate, but sufficiently high to encourage the absorption of the mineral principles contained in the water. The degree of mineralisation must also vary with different patients. When taken under suitable conditions, the baths exercise a sedative, tonic, solvent, resolutive, and restorative action.

The same remarks apply to the douches.

The special use of the springs of Salins, as of all waters containing strong solutions of chloride of sodium and bromo-iodides, is in the treatment of scrofula, from simple *lymphatism* up to its most serious manifestations, such as necrosis, caries, white swellings, coxalgia, and spinal curvature. These waters are also indicated in scrofulous phthisis, which is beneficially affected by them, and often, even cured.

The Salins baths are superior to sea-bathing in two respects; first, that the water is drinkable, and second that its mineralisation can be graduated, while that of the sea is of course constant.

A residence at Salins is very enjoyable. The Establishment possesses a splendid casino and garden; and the town contains (among other objects of interest) the two forts, the church of St André (which ranks as an historical monument), the library, the fountain in the *Place d'Armes*, the work of *Devoge*, an eminent sculptor of the last century, and the *Barbarine* garden. In the neighbourhood, pleasant excursions may be made to the *Pont du Diable*, the *Lison* springs, the *Grottes des Planches*, *Mont Poupet*, the *Bout du Monde*, and the magnificent pine forests which surround the town.

URIAGE

(*Isère*)

MIXED SULPHURO-CHLORINATED SPRING (TEPID), AND FERRUGINOUS SPRING.

URIAGE is a small town with a population of 2500, situated about eight miles from Grenoble, in a

picturesque valley at the foot of the eastern slope of
the Alps of Dauphiné, the long and beautiful range of
mountains through which one passes on the way to
Grenoble, and from Grenoble to the Thermal Station.

Uriage is neither a town—properly speaking—nor
a village; it is a collection of handsome buildings, all
designed for the service of the baths and the bathers,
and all perfectly adapted to their object.

There are two mineral springs; one is ferruginous,
and is drunk in cases where iron is indicated; the
other is both saline and sulphúrous. It is to this
second spring that Uriage owes its reputation, and it
is the only one which calls for notice.

The baths of Uriage were known to the ancients, as
is clear from the numerous remains of Roman Thermæ
discovered in the process of boring for the springs, and
excavating for the foundations of the present Establishment. This latter is one of the handsomest and most
complete in France. It contains more than ninety
large and convenient bath-rooms, some with two,
three, or even four baths, but the gréater number
with only one. Each bath is provided with four taps,
two for mineral water, at its natural temperature, and
heated; and two for ordinary water, cold and hot.
The arrangements for douches are also very complete,
consisting of ten apartments for general and local
douches of every variety, and two for vapour and
Russian baths. There are also a *buvette* or drinking-
fountain, and two inhaling rooms, one for vapour and
gas, the other for pulverised water and gas.

The spring which supplies the Establishment is the same that once fed the Roman baths. It issues from the schists by several fissures, and flowing along a conduit formed of fir trunks hollowed out, protected by a tunnel of masonry nearly 1000 feet long, is conducted into a vast reservoir, which contains about 265,000 gallons of water. From thence it is conveyed to the different parts of the Establishment.

The waters of this spring have a normal temperature of 27° C. (80° F.) They are perfectly clear and colourless when they spring from the rocky fissures, but grow turbid on contact with the air, in consequence of the precipitation of a part of the sulphur which they contain. Their smell is distinctly sulphurous, and their taste, at first resembling gall, becomes very salt, and somewhat bitter.

CHEMICAL ANALYSIS

WATER 1000 grammes (1 litre).	Cubic centimètres.	Grammes.
Nitrogen	19·5	
Carbonic acid, free	3·2 or	0·0062
Sulphuretted hydrogen	7·3443 or	0·0113
Chloride of potassium		0·4008
„ sodium		6·0569
„ lithium		0·0078
Sulphate of lime		1·5205
„ magnesia		0·6048
„ soda		1·1875
Bicarbonate of soda		0·0555
Arseniate of soda		0·0021
Silica .		0·0790
Traces of chloride of rubidium, iodide of sodium, hyposulphite of soda, sulphide of iron, and organic matter.—*Lefort*.		10·4262

From the preceding analysis, it appears that the predominant mineral principles of the Uriage waters are chloride of sodium, sulphate of lime, sulphate of soda, and free sulphuretted hydrogen. They are, therefore, sulphuro-chlorinated mixed.

When in doses of from one to two glasses per day, the Uriage waters are aperient; they gently stimulate the mucous membrane of the digestive organs, excite a slight thirst, and cause increased activity in the functions of the stomach and intestines. In larger doses, of from three to six glasses, they readily act as a purgative, causing abundant evacuations, without griping, and generally producing their action within two or three hours.

The baths at Uriage are tonic and strengthening; such is their usual and characteristic quality, but the effects vary according to the temperature at which they are administered. When used cold, and for a short time only, they have a steady tonic action, and produce a marked sedative effect over the nervous system. Thus taken, they do not in any way induce absorption. But if used luke-warm, i.e. at a temperature of 32° C. (90 F.), they favour cutaneous absorption, and cause an abundant excretion of urine. With some persons they produce a more or less marked nervous excitement, which passes off at the end of a few days; with others, their tonic and reviving action is shown by an increase of strength, a greater muscular activity, and a very appreciable sense of general comfort.

All that has been said of baths applies equally to douches, the use of which is always combined with *massage* (shampooing), as practised at Aix-les-Bains. The action of the air of the inhaling rooms is chiefly due to the presence of sulphuretted hydrogen gas. Under its influence, the circulation becomes more active, the pulse is strengthened and accelerated, the skin becomes warm and moist, and the pulmonary organs are at first excited, although afterwards a marked tendency to sedative reaction is observed.

In the first rank of the diseases, properly so-called, for which the waters of Uriage are especially suited, stands scrofula in all its manifestations, whether they simply affect the skin (tubercle, ulcerations), or the mucous membranes (chronic ophthalmia, coryza, ozæna, otorrhœa), or the osseous system (rickets, caries, necrosis). They have even conquered mesenteric swelling, and strumous congestion of the pulmonary lymphatic glands, thus stopping the development of consumption.

It has been said that these waters are particularly well suited to cutaneous affections, especially those of the eczematous variety. While waters which are simply sulphurous, like those of Barèges or Luchon, stimulate the skin strongly at the commencement of the treatment, those of Uriage have at once a soothing effect, which must, without doubt, be attributed to the sedative and somewhat astringent action of the chlorides. Some credit must also be given to the water drunk, the laxative effect of which prevents a too great congestion of the skin. It is only after the first few baths that the surface becomes irritated, and even then the irritation is usually slight, and lasts but a short time.

The Uriage waters do not suit acute affections, advanced organic degeneration, or persons suffering from cancerous or tubercular diathesis in its final stage. They are also contra-indicated, and would be injurious, in sanguineous and plethoric temperaments; in diseases of the heart or large blood-vessels, and in paralysis of recent date.

Lastly, it may be mentioned that Uriage possesses a ferruginous spring, discovered in 1845, the use of which may often be advantageously combined with that of the sulphuro-chlorinated spring, as most of the maladies beneficially affected by the one are equally influenced by the other.

Uriage has a very good casino, with two large and handsome ball-rooms, reading and conversation *salons*, card, billiard, and smoking rooms, and a concert hall.

The environs are very picturesque and interesting, and numerous excursions may be made to places of interest. Amongst these are the Château of Uriage, with its fine picture gallery, including originals of Paul Veronese, Rubens, Teniers, van Ostade, Correggio; the colossal allegorical statue of the Alps by Sappay; the church of *St Firmin*; the villages of *Saint-Martin*, *Saint-Nizier*, and *Villeneuve*; the mountains known as *des Quatre-Seigneurs* and *de Combeloup*, &c.

ALLEVARD

(*Isère*)

CALCAREO-SULPHURETTED WARM SPRING

ALLEVARD is a small town of 5000 inhabitants, situated on the banks of an impetuous torrent named the *Bréda*. It is about 1480 feet above the sea level, and twenty-five miles from Grenoble, in the midst of grand mountains. It has long been reputed for its steel works. Tourists visit the place to admire the mountains and glaciers, and a wild valley, 7850 feet above the sea, in which there is a chain of seven lakes. The waters, which had been known for a long time to the inhabitants under the name of "the black waters," have only been generally used for about forty years. They are calcareo-sulphuretted, with a temperature of 24° C. (75° F.) at the emerging point. There is only one spring, which rises from the black limestone that covers the country, discharging about 882,000 gallons of water in twenty-four hours. It is received in a well, whence it is raised by a pump driven by the torrent, and conveyed to the *buvette*, to a reservoir, and to the boiler, in which it is heated in such a manner as to prevent any alteration in its chemical properties. The Establishment is situated in a large park, containing lovely walks and shady woods. It has every arrangement for the special employment of its sulphurous waters, which are unusually rich in sulphuretted hydrogen and carbonic acid.

There are seven large rooms for cold gaseous inhalations, two for vapour inhaling, and numerous rooms for pharyngeal douches, and pulverised water, for affections of the larynx; several "grand" douches, some ascending ones, and a most complete set of apparatus for hydrotherapeutics. There is also a large newly-constructed hall specially reserved for "gargling," and communicating with the new *buvette*; the whole forming a complete and varied Establishment.

The different departments are connected by a fine glazed gallery, which forms a pleasant promenade in wet weather.

CHEMICAL ANALYSIS

WATER 1000 grammes (1 litre).	ANHYDROUS SALTS.	CRYSTALLISED SALTS.
	Cubic cent.	
Sulphuretted hydrogen . . .	24·75	
Carbonic acid of bicarbonates and free . . .	97·00	
Nitrogen	41·00	
	Grammes.	Grammes.
Carbonate of lime . . .	0·305	0·305
„ magnesia . .	0·010	0·015
Sulphate of soda . . .	0·535	1·211
„ magnesia . .	0·523	1·065
„ lime . .	0·298	0·375
Chloride of sodium . .	0·503	0·503
„ magnesium . .	0·061	0·061
Silicic acid . . .	0·005	0·005
	2·240	3·139

Traces of carbonate of iron, sulphate of alumina, bituminous matter, glairine, and chloride of aluminium.—*Dupasquier.*

At the spring, the water has an opaline tint, due to

the bubbling up of the gas. Colourless when still, it becomes cloudy if agitated, and clears again, after leaving a deposit of sulphurate and carbonate of lime. Seen in a mass in the bath it has a greenish shade, and like that of *la Reine* at Luchon, shows the phenomenon of *whitening*, owing to the decomposition of its hydrosulphuric acid.

The water of Allevard has a stronger smell and fainter taste than that of Enghien; the flavour is fresh, bitter, somewhat salt and astringent, and patients easily accustom themselves to it. It raises the general constitutional tone, stimulates the circulation and the functions of nutrition, and acts especially upon the mucous membranes and the skin.

The complaints to which these springs are applicable belong to the same class that are usually treated with sulphurous waters in general, *i.e.* diseases of the skin and of the respiratory organs. It has been said that Allevard water, suitably administered, is capable of arresting the progress of phthisis, of assisting in the expulsion of tubercle, and of bringing alleviation even in the hectic stage. It is also recommended, in the form of drink and inhalation, in cases of affections of the larynx, in aphonia, pharyngeal granulations, bronchial catarrh, chronic pleurisy, and asthma; externally in baths and douches, for rheumatism, leucorrhœa, spermatorrhœa, &c.

The establishment is open from the 1st of June to the end of September.

The neighbourhood of Allevard possesses several

objects of interest, to which excursions or walks may
be made. The blast furnaces of the steel works can
be also visited, and the ruins of a feudal castle ;
le mont Brame-Farine, from the top of which the *Mont
Blanc* and the Dauphiné Alps may be seen; the
Chartreuse de Saint-Guyon, the *Château Bayard, le
Fort Barraux, le Mont des Sept-Laux,* &c. Guides
are to be found at Allevard who provide horses,
carriages, and donkeys at prices which are fixed, each
year, by an official tariff.

SAVOY

AIX-LES-BAINS

(*Savoie*)

SODIO-CALCAREO-SULPHURETTED HOT SPRINGS

Aix-les-Bains is a pretty little town, which became
French by the annexation of Savoy in 1860. It has
a permanent population of 4000 ; but the number of
visitors more than doubles it during the season.
Situated on the east of the valley of *Aix,* on the
slope of a picturesque hill, at a height of over 800
feet above the level of the sea, this thermal station
enjoys so healthy a climate, that neither cretinism nor
goître (so sadly common in the other Alpine valleys)
is to be met with there.

The waters of Aix were known to, and frequented

by the ancients; and the Emperor Gratian built there baths, which were called after him *Aquæ Gratianæ*, and of which the fine remains can still be admired, especially those of the vaporarium.

The existing Establishment was founded in the year 1776, and had been several times enlarged and altered, until it became the property of the State, in 1860; and since then an unsparing expenditure has so increased its accommodation and perfected its arrangements that it is at present incontestably one of the most remarkable in Europe. It contains sixty-four bath rooms, six public baths with running water (of which two are swimming baths), two small for family use, and two of medium size; fifty large general douches of all kinds, six local douches, and several ascending douches in the bathrooms; two vapour baths, two inhalation rooms, two pulverisation rooms, and one *buvette*.

There are two springs, quite distinct from each other, and giving altogether a total of 1,322,760 gallons per day, a quantity sufficient to supply 1200 baths, 2000 douches, and more than 200 inhalations. The water is sulphurous and mixed bicarbonated; that of the spring called *Source de Soufre* has a temperature of 45° C. (113° F.); while the temperature of the other, called *Source de St Paul*, is 46·5° C. (115·7° F.) This latter spring bears also the old name of the "Alum Spring," which is a totally misleading designation, as it contains no alum.

The waters of both springs are clear and colourless.

They smell (especially that of the " Sulphur " spring) strongly of rotten eggs, and have a sweetish taste, slightly nauseating, and inducing risings of a very distinctly sulphurous kind. They are used for drinking and inhaling; in ordinary and vapour baths, in douches, and with *massage* or shampooing. Their physiological effects are various. They are tonic and restorative, stimulate the nervous system, and have a very marked action upon the skin, and on the mucous membranes of the digestive organs, and the urinary passages.

CHEMICAL ANALYSIS

WATER 1000 grammes (1 litre).	SOURCE DE SOUFRE.	SOURCE DE ST. PAUL (OU D'ALUM).
	Litre.	Litre.
Nitrogen	0·0252	0·0631
Carbonic acid, free	0·0130	0·0067
Sulphuretted Hydrogen, free	0·0267	—
Oxygen	—	0·0041
	Grammes.	Grammes.
Silicic acid	0·00500	0 00430
Phosphate of alumina and lime } Fluoride of calcium	0·00249	0·00260
Carbonate of lime	0·14850	0·18100
„ magnesia	0·02587	0·01980
Bicarbonate of iron	0·00886	0·00936
Sulphate of soda	0·09602	0·04240
„ lime	0·01600	0·01500
„ magnesia	0·03527	0·03100
„ alumina	0·05480	0·06200
Chloride of sodium	0·00792	0·01400
„ magnesium	0·01721	0·02200
Loss	0·01200	0·00724
	0·43000	0·41070

Traces of bicarbonate of strontia, sulphate of iron, alkaline iodides and glairine.

The complaints against which the Aix waters have a real efficacy are numerous. In the first rank must be placed rheumatism, and especially chronic rheumatism of the joints, muscles, and viscera, which often yields to a combination of baths, douches, and shampooing (*massage*).

The manner in which this treatment is carried out at Aix deserves description. An invalid, for instance, who is unable to walk, would be carried from his bed to the Establishment in a closed *chaise-à-porteurs*, accompanied by an attendant (called a *sécheur*, or drying-man) carrying his bath towels, linen, &c. Arrived at the douche room, he is seated on a stool between two *masseurs* or shampooers, who knead him, rub him, and work his joints, according to the prescriptions and directions of the doctor. At the same time they direct on him two jets of mineral water, one—the warmest and of the highest pressure—upon his limbs; the other —cooler and of very weak force—upon his neck, back, and chest.

During all the time that the douche lasts, the patient is therefore bathed with a perfect stream of mineral water at a maximum temperature of 45° C. (113° F.), and a minimum of 16° C. (60·8° F.) He is then dried, carefully wrapped in woollen coverings, and carried back in the *chaise-à-porteurs* to his bed, which has in the meantime been warmed, and where he can continue (if so directed) the process of sudation which had begun.

After rheumatism come diseases of the skin, such

as herpetic eruptions, prurigo, *couperose*, eczema,
psoriasis, &c.; scrofulous affections, rheumatoid arthritis, white swellings of joints, chronic ulcers,
fistulous openings, false anchyloses, and chronic bone
diseases; the sequelæ of dislocations, fractures, and
unhealed sores resulting from traumatic causes. The
waters of Aix have also been recommended as
efficacious against syphilitic affections, especially
those of a squamous and tuberculous nature. Among
diseases of women may be noted chlorosis, hypertrophy
of the neck of the uterus, erosions, granulations, and
ulcers, as well as displacements of that organ. Very
obstinate cases of sciatica are also said to have been
cured at Aix.

There are other complaints which, far from being
alleviated by these waters, are aggravated by them.
Among these are all acute affections, cachectic diseases,
predisposition to cerebral congestion; nearly all cancerous degenerations; consumption, and diseases of
the heart.

The season at Aix-les-Bains begins on the 1st of
May, and ends in the month of October. The course
of treatment usually takes about twenty-one days.

Visitors will find all kinds of interesting amusements
at their disposal. The Casino is handsome and very
well organised; it contains concert, ball, and reading
rooms; card and billiard rooms, café, restaurant,
gardens and covered galleries; in short, nothing is
wanting. In the town the chief objects of interest
are the tunnel where the "Alum Spring" rises, which

forms the principal entrance to the caves of *St. Paul ;* the ruins of the *Roman baths*, the arch of *Campanus*, the temple of *Diana*, and the *Mollard* garden.

Many delightful excursions can be made in the neighbourhood.

MARLIOZ

(*Savoie*)

SODIO-SULPHURETTED COLD SPRINGS

MARLIOZ is a little hamlet, at a few minutes' walk from Aix-les-Bains, which owes its existence to its sodio-sulphuretted springs. These are three in number, and are known by the names of the *Source Esculape, Source Adelaïde*, and *Source Bonjean*, all of practically the same degree of mineralisation.

The Thermal Establishment (which was built about twelve years ago) is exclusively devoted to the inhalation and pulverisation of the sulphurous water, these two forms of administration being specially used at Marlioz. Baths are almost unknown there, and only a few persons drink the waters.

These are clear and colourless, and show bubbles of gas escaping at intervals. They deposit inside the basins an abundant slimy sediment; are oily to the touch, and have a smell and taste strongly resembling that of gall. They grow turbid on contact with the air, and, curious to relate, when they are allowed to become still in a glass, they decompose, and the sulphur which they contain can be seen settling at the bottom. The arrangements of the Establishment and the chemical composition of the Marlioz waters are sufficient to indicate their action on the respiratory organs. In this respect they resemble those of Eaux-Bonnes, Saint-Honoré, &c., and are suited to the same class of complaints.

In consequence of the nearness of Aix, patients who take the Marlioz waters, live at the former place, and either walk or drive to the latter every day, in order to take their inhalation or pulverisation.

CHALLES

(*Savoie*)

SODIO-SULPHURETTED COLD SPRING, WITH CHLORIDES AND IODIDES.

CHALLES is situated at about three miles from Chambéry. Its mineral spring was only discovered in the year 1841. The water is cold and transparent, and every now and then releases little bubbles of nitrogenous gas. Its taste has a slight bitterness, to which the palate soon grows accustomed, and it has almost no smell.

Challes water is used both as a drink and for baths. The latter used to be taken at Chambéry, but an Establishment has now been built at Challes itself, close to the spring. Besides very complete bathing arrangements, this Establishment has two inhalation rooms, two for pulverisation, and a perfect arsenal of hydrotherapeutic apparatus.

The water of Challes is quite unique in its composition, its mineralisation being extremely peculiar. It contains so large a quantity of sulphur that it might almost be called an essence of sulphur water; yet it is also impregnated with iodides and bromides in larger proportion than any other known sulphurous water. This is shown by the analyses of Dr Garrigou, as follows:

CHEMICAL ANALYSIS

WATER 1000 grammes (1 litre).	
	Grammes.
Chloride of magnesium	0·0100
„ sodium	0·0814
Bromide of sodium	0·0100
Iodide of potassium	0·0099
Sulphide of sodium	0·2950
„ iron and manganese	0·0015
Carbonate of soda	0·1377
Sulphate of soda } „ lime }	0·0730
Silicate of soda .	0·0410
Bicarbonate of lime	0·0430
„ magnesia .	0·0300
„ strontia .	0·0010
Phosphate of alumina and lime } Silicate of alumina and lime }	0·0580
Rudimentary glairine	0·0221
Loss	0·0325
Traces of nitrogenous organic matter, free soda, and slight traces of nitrogen.—*O. Henry.*	0·8461

It appears from this that the waters of Challes contain a considerable quantity of sulphides, and a considerable proportion of iodide and bromide of sodium. Now, iodine and bromine are specifics for scrofula; and sulphur, besides its general dynamic and exciting power, has a special action upon the skin. Consequently, Dr Bazin strongly recommends these waters in complaints arising from scrofula, even in the most inveterate cases. They are also very useful in goître, ganglionic swellings, ulcers of long standing; in cases of caries and necrosis; tertiary syphilis, chronic rheumatism, atonic gout, and in general all chronic inflammatory affections. It is also said that excellent results are being constantly obtained in the treatment of ozæna.

Until lately there was no accommodation at Challes for taking the waters on the spot. But the " Château," an old seigneurial residence, has recently been transformed into a very comfortable hotel for the use of visitors to the springs. It is picturesquely situated on a height commanding a magnificent view, and surrounded by trees, so that it is an unusually attractive place of residence. Various interesting excursions can be made in the neighbourhood, among which may be mentioned those to the *Château de Coynin;* the *Charmettes,* immortalised by Jean Jacques Rousseau and Madame de Warens; the *Bout-du-Monde,* a picturesque ravine ending in a wall of precipitous rocks, over which falls a cascade, &c.

LA BAUCHE

(*Savoie*)

FERRUGINOUS, CARBONATED AND CRENATED,
COLD SPRINGS.

LA BAUCHE is situated in the centre of one of the most fertile valleys of Savoy, on the slope of the *Signal* mountain, from whence the eye ranges over a marvellous panorama, and near a magnificent pine forest. Its waters were already celebrated in the time of the Romans, as is shown by the remains of all kinds found near the springs, and the traces of ancient roads dating from that period.

From the analyses of MM. Callond and Abbène, it appears that these waters contain gram. 0·173 per litre of crenate and bicarbonate of protoxide of iron, a larger quantity than is contained by the richest of other waters of the same class.

La Bauche possesses, besides its ferruginous springs of the first rank, a fine stream of water, which coming from a height of at least 260 feet, would be fully capable of supplying a hydropathic Establishment.

The waters bear exportation perfectly. They are remarkably digestible, and make excellent table-waters, when mixed with wine, which they do not decompose. They will be found to exercise a generally tonic and strengthening effect upon weak constitutions.

SALINS-MOUTIERS

(*Savoie*)

SODIO-CHLORINATED, WARM SPRINGS.

SALINS, which was formerly an important town, is at the present time only a modest village, about a mile from Moutiers, in a narrow valley traversed by the River *Doron*.

The thermal springs, of chloride of sodium, have been known from time immemorial, and used by the inhabitants for making salt. Their therapeutical application only dates from 1838. Their temperature is 36° C. (97° F.); and they have a flow of 106,000 cubic feet of water per day. It is clear and limpid in a glass, but if seen in quantities in the reservoirs it has a clear orange colour, due to the ferruginous deposits on the sides; and numerous bubbles of carbonic acid gas continually rise to the surface. Rough to the touch, without any perceptible smell, it has a distinct saline flavour, with a slightly bitter taste. Nevertheless, notwithstanding this very pronounced taste, it is not disagreeable to drink.

It is easy to see from their chemical composition that the waters of Salins are tonic, alterative, resolutive, and restorative. They are used, as at Salins du Jura and at Salies-de-Béarn, for drinking, and also for baths and douches; either pure, or with the addition of a certain proportion of concentrated water, according to the requirements of the case.

CHEMICAL ANALYSIS

WATER 1000 grammes (1 litre).	Grammes.
Carbonate of lime	1·005
Sulphate of lime	1·392
„ magnesia	0·752
„ soda	0·641
Chloride of sodium	11·317
Insoluble residue	0·036
Traces of iodine, iron, arsenic and organic matter.	
—*Bouis.*	15·143

The diseases successfully treated at Salins are glandular swellings, caries, fistula, indolent ulcers, and, in a word, all the series of scrofulous affections.

BRIDES-LES-BAINS

(*Savoie*)

SULPHATO-CHLORINATED AND BICARBONATED SPRING

BRIDES-LES-BAINS is a pretty little village, situated about three miles from Moutiers, at about 1870 feet above the level of the sea. It stands in a charming valley, surrounded by high mountains covered with glaciers.

The mineral waters, which contain sulphates, chlorides, and bicarbonates, have a temperature of 30° C. (86° F.) ; they spring from a single source on the

left bank of the *Doron*, a few steps from the Thermal
Establishment, which is very comfortably arranged.
They are limpid, soft to the touch, and have a slightly
acid and styptic taste. Quantities of bubbles of
carbonic acid gas are given off, and make them sparkle
like artificial aërated waters.

CHEMICAL ANALYSIS

WATER 1000 grammes (1 litre).	Grammes.
Carbonic acid, free	0·600
Sulphate of lime	2·350
„ soda	1·031
„ magnesia	0·700
Chloride of sodium	1·222
Carbonate of lime	0·325
„ and oxide of iron . . .	0·016
Silica	0·042
Traces of iodine, arsenic, and phosphates.—*Gobley.*	6·286

The waters are administered internally, and externally as baths and douches. They are tonic, laxative,
and restorative, according to the quantity taken; a
somewhat large dose is distinctly purgative without
fatiguing the stomach. The diseases in which Dr Laissus,
the inspector, has prescribed these waters with the
greatest success are: flatulence, dyspepsia, obstinate
constipation, catarrh of the urinary passages, certain
uterine affections; but specially congestion of the liver,
for which particular malady they rival Vichy and
Carlsbad. It is also said that they are a sovereign
remedy for obesity.

Unfortunately this Station is too difficult of access for it to be utilised to the extent warranted by the value of its waters.

EVIAN-LES-BAINS

(Savoie)

SODIO-CALCAREO-BICARBONATED SPRINGS, (FAINTLY MINERALISED).

EVIAN is a little town, with a population of 2500, on the Savoy side of the Lake of Geneva. It is built in the shape of an amphitheatre, close to the water's edge, and looks across to Lausanne upon the opposite shore. The climate is mild, the air is healthy, and the situation charming, commanding a magnificent view.

Its mineral springs, which contain bicarbonate of soda and lime, supply two Establishments : the *Cachat*, lately restored, and the *Bonne-vie*, which answers all the requirements of modern hydrotherapeutics. They are six in number, the two principal bearing the same names as the two Establishments, and their identity of composition proves that they emanate from the same subterranean reservoir. The water is cold, its temperature being only 12° C. (53·6 F.), colourless, alkaline, without any smell, and of a very agreeable taste.

WATER 1000 grammes (1 litre).				SOURCE BONNEVIE.
				Grammes.
Bicarbonate of lime	.	.	.	0·2210
„ magnesia	.	.	.	0·0150
„ soda	.	.	.	0·0200
„ potash	.	.	.	0·0070
Phosphate of soda	.	.	.	0·0017
Carbonic acid, free	.	.	.	0·0970
Traces of alkaline chlorides.—*École des Mines.*				0·3617

The Evian waters are drunk, and used for baths and douches. As shown by the analysis, they are very faintly mineralised, and nothing in their chemical composition indicates their well-known therapeutical value. They agree perfectly with the stomach, and act as an excellent diuretic in cases of gravel, enlargements of the prostate, and catarrhal affections of the bladder and kidneys, by the species of irrigation which they produce in the interior of these organs. If there is irritability of the urinary passages (caused by the presence of a calculus, or after lithotritic operations), they are to be preferred to those of Vichy, Vittel, or Contrexéville, which in such cases would be much too exciting.

Chronic affections of the digestive organs, such as dyspepsia, gastralgia accompanied by acid eructations, pyrosis, chronic irritation of the intestines, chronic diarrhœa, obstinate constipation, hæmorrhoids, &c., are all treated with benefit at Evian. Patients suffering from the irritability which accompanies most forms of chronic neuralgia, have been relieved from it after using the waters for some days.

They may be recommended as the most perfect type of table-waters, in consequence of their complete freedom from sulphate of lime, their clearness and agreeable taste.

Visitors will find various interesting objects for walks in the neighbourhood, among which may be enumerated : the *Château de Neucelles* with its great chesnut-tree, which is more than forty-six feet in circumference, and can contain several persons at a time inside its hollow trunk ; the ruins of the *Château de Maxilly*, the Grotto of Jean Jacques Rousseau at *la Meillerie*, and the Abbey of *Ripaille*. Those who are fond of fishing will find sport in catching the trout that abound in Lake Leman.

AMPHION-LES-BAINS

(*Savoie*)

ALKALINE AND FERRUGINOUS COLD SPRINGS

AT twenty minutes' walk from Evian, and on the very shores of the lake, stands the Thermal Establishment of Amphion-les-Bains, supplied by four springs, one of which is ferruginous bicarbonated, while three are alkaline. The latter resemble in chemical composition and medicinal properties those of the *Cachat* and *Bonne-vie* at Evian. The former contains a

notable quantity of oxide of iron, in conjunction with
alkaline carbonates, which facilitate its assimilation.
The temperature is 10° C. (50° F.)   The water of
this spring is prescribed with success whenever the
use of iron is indicated.   It is so abundant as to be
used not only for drinking, but also in public and
private baths.

Amphion has one of the most magnificent points of
view in Europe.   From the top of the hill which
shelters it—can be seen the grandest and most varied
panorama that it is possible to imagine.

THE springs of this region belong to two classes only, the sulphurous and the ferruginous; but although they are not numerous, they yet rank among the most important in France, both on account of their medicinal value, and of the geographical position and completeness of the Establishments in which they are administered.

ENGHIEN

(*Seine-et-Oise*)

CALCAREO-SULPHURETTED COLD SPRINGS

ENGHIEN is a small town of modern date, situated at the gates of Paris. "A charming situation, a lake in proportion to the landscape, elegant houses, admirably laid out gardens; everywhere flowers, trees, walks, shade, beautiful effects of light, something which reminds one of the happiest of lands, and the most favoured of climates." Such is the description which has been given of this thermal locality, and it is more than justified by the facts.

The Enghien springs are sulphureous. The sulphur which they contain is in the form of sulphuret of lime,

and abounds still more in hydrosulphuric acid gas.
The proportion of sulphur in the water greatly exceeds
that in the waters of the Pyrenees. It is true that on
the other hand the latter are naturally warm, whereas
those of Enghien are cold,—and that the springs of the
Pyrenees are very rich in *barégine,* while those of
Enghien do not contain a particle of this substance.

The principal Establishment in which the Enghien
waters are used contains more than 100 baths; all are
provided with three taps, one of cold sulphurous water,
and the two others of ordinary water, hot and cold.
Each bath is furnished with douches of high and low
pressure, which can be used either with or without
the baths; they are ascending or descending, utero-
vaginal, and rectal.

Besides these douches connected with the baths, there
are others completely separate, comprising all possible
varieties of force and direction; and on the whole, the
completeness of the arrangements and the diversity of
the apparata render this Hydrotherapeutic Establishment one of the most remarkable in France.

The second or " Small Establishment," situated a few
hundred yards from the first, is of more recent construction. The management decided upon erecting it,
when three new springs were discovered in 1865. Its
hydropathic arrangements are almost the same as those
of the larger Establishment, but on a more modest
scale, so as to render it available to persons whose
resources are too slender to meet the expense of the
latter.

These two Establishments are supplied by several springs belonging to the class of calcareo-sulphuretted waters. The principal are those called the *Source du Roi, Source Deyeux, Source Péligot, Source Bruland, Source de la Pécherie, Source du Lac, Source des Roses,* and *Source du Nord.* The water of all is clear and colourless, of a smell and taste more or less hepatic according to the spring, and deposits a yellowish sediment on the lining of the basin. Their chemical composition is nearly identical, and varies only in the proportions of the elements, as has been shown by analyses made at different times.

The following recent analysis gives the average composition of the *Sources du Roi, Deyeux, du Nord,* and *du Lac.*

CHEMICAL ANALYSIS

WATER 1000 grammes (1 litre).	
	Grammes.
Carbonic acid, free	0·1419
Sulphuretted hydrogen	0·0453
Nitrogenous organic matter . . .	0·1390
Sulphate of potash	0·0140
„ soda	0·0176
„ alumina	0·0247
„ magnesia	0·0758
„ lime	0·2493
Silicate of magnesia	0·0254
„ lime	0·0580
Carbonate of magnesia	0·0083
„ lime	0·2773
	0·7508

The action of the Enghien waters is analogous to
that of the sodio-sulphuretted springs of the Pyrenees;
that is to say, they are stimulating, tonic, and restorative. Their stimulating action is chiefly exercised
upon the skin, and the mucous membrane of the air-
passages. They can be taken in all forms, but are
principally used as baths, drinks, or inhalations—
either in a state of vapour or pulverisation. Used in
baths, they often excite *la poussée* (thermal fever), and
sometimes so energetically that the sulphurous water
has to be tempered by dilution with ordinary water.
Used as a drink, they cause to some persons a feeling
of weight in the stomach. In this case they must be
taken mixed with a little milk, for which purpose
cows', asses', and goats' milk is always to be had at
the *buvette*. Finally, the Enghien waters, inhaled as
vapour or spray, act directly on the mucous membrane
of the air passages, and stimulate or modify its
secretion. They also have a sedative action on the
circulation, and have the property of slackening the
pulsations of the heart.

The class of diseases to be found in the largest
numbers at Enghien are affections of the respiratory
organs, and skin diseases.

The affections of the respiratory organs which can
be cured or alleviated by the use of the Enghien
waters, either drunk or inhaled as vapour or spray,
are, in order of frequency: bronchitis, laryngitis, catarrhal pharyngitis, asthma (both when complicated with
herpetism and when coexisting with a rheumatoid

condition) ; and also phthisis both in the first and
second stages, if the constitution of the patient is
scrofulous or lymphatic.

The skin diseases for which the Enghien waters
furnish a really legitimate and efficacious remedy
are those connected with the herpetic diathesis.

Residence at Enghien is extremely agreeable; and
regattas on the lake, concerts, balls, the casino; promenades to the *Bois Jacques,* to *Montmorency, Saint-
Gratien, Sannois,* the *Hermitage, &c.,* can amuse
patients during the process of their " cure."

PIERREFONDS

(*Oise*)

CALCAREO-SULPHURETTED COLD SPRINGS, AND FERRU-
GINOUS BICARBONATED AND ARSENICAL SPRING.

PIERREFONDS is a town of 1700 inhabitants, situated
at the eastern end of the forest of Compiègne. It
was formerly celebrated for the magnificent ruins of
its fortress, which recalled those of Heidelberg ; but
the building has now been completely restored by
Violet-le-Duc. The town lies at the foot of the hill
upon which the castle is built, near a little lake.

The mineral springs, which were only discovered in
1845, belong to the two classes of cold calcareo-sulphuretted and ferruginous waters.

The sulphurous springs are collected in a reservoir constructed in the lake itself. This reservoir supplies the Thermal Establishment, which comprises twenty-one bath-rooms, eight douche-rooms, one ascending douche, and a large room for inhaling the sulphureous water in a state of pulverisation. The *buvette* is placed in a little pavilion at the extremity of the park belonging to the company, and is supplied by a special spring. These waters are perfectly clear when they issue from the springs, but grow turbid on contact with the air, and become first blueish and semi-opaque, then white and milky, and ultimately deposit sulphur on the walls of the basins. Their smell is like that of addled eggs, and their flavour, though slightly resembling gall, leaves behind no disagreeable after taste.

CHEMICAL ANALYSIS

WATER 1000 grammes (1 litre).	Grammes.
Sulphuretted hydrogen, free . .	0·0022
Sulphide of calcium . . .	0·0156
Sulphate of lime ⎱ soda ⎰ . .	0·0260
Bicarbonate of lime ⎱ magnesia ⎰ .	0·2100
Chlorides of sodium and magnesium .	0·0220
Silica and alumina ⎱ Salts of potash ⎱ Organic matter ⎰ . .	0·0500
Traces of carbonic acid and nitrogen.	0·3358

The waters of Pierrefonds are principally used for

drinking and pulverisation. As a drink the maximum dose is two glasses, morning and evening. The treatment with pulverised water originated at Pierrefonds, and the credit of its conception is due to the distinguished Inspector, Dr Sales-Girons. It is carried on in a special room, admirably arranged and fitted up for the purpose, during periods ordinarily of forty-five minutes each. Baths and douches are also used at Pierrefonds, but not on so large a scale.

The therapeutical properties of the Pierrefonds waters are those of sulphureous waters in general. Nevertheless, the diseases most generally found under treatment there, are (as at Eaux Bonnes) affections of the respiratory organs, such as laryngitis, pharyngitis, bronchial and pulmonary catarrhs, asthma, and phthisis.

The ferruginous spring is situated at a few paces from the sulphurous *buvette,* and covered by a little pavilion. The iron which it contains is in the form of bicarbonate and arseniate of iron, as has been shown by chemical analysis.

After affections of the respiratory organs, those which are most frequently met with at Pierrefonds are chlorosis, anæmia, amenorrhœa, and dysmenorrhœa (which are beneficially affected by the combined use of the two different waters); rheumatism; and certain diseases of the skin.

Life at Pierrefonds is very pleasant. The morning is devoted to the treatment, the afternoon to walking, and the evening to amusement. The theatre, concerts, dances, charades, play, reading, and conversation, can all be indulged in according to the various tastes of the patients,—while the walks are so numerous and varied that a new one could be taken every day for a month.

Among the objects of interest in the neighbourhood are the magnificent *Château de Pierrefonds*, the *Château du Prieuré*, the delicious valley *de la Folie*, the pond at *Batigny*, the *Château de Vez*, the *Château d'Offémont*, the pheasantry, and the splendid forest of Compiègne.

FORGES-LES-EAUX

(*Seine-Inférieure*)

FERRUGINOUS COLD SPRINGS

FORGES, situated in the Department of the Seine-Inférieure, is a small place of 1600 inhabitants, built on the slope of a hill which overlooks the smiling and fertile valley of Bray. Its mineral waters, which had been known from the fourteenth century, owed their celebrity to a visit paid to them in the year 1632, by Louis XIII, Anne of Austria, and Cardinal Richelieu. On September 16th, 1638, after twenty-three years of sterility, Anne of Austria brought into the world the infant who afterwards became Louis XIV. Later on, in the year 1772, the Duchess of Chartres, subsequently Duchess of Orleans, who had been married three years, and was still childless, also came to take the waters of

Forges, and on October 6th of the following year, she gave birth to a son who became Louis-Philippe.

At the present day, Forges is by no means so much in fashion as it was in the seventeenth century. Nevertheless, the number of its frequenters has of late largely increased, thanks to the energy and intelligence of the new management, which has substituted for the old Establishment a modern building better adapted to the purposes for which it is required.

The waters of Forges are ferruginous. They flow from three springs, named respectively the *Reinette*, the *Royale*, and the *Cardinale*. There is a separate *buvette* for each spring, and in the Thermal Establishment everything is to be found that is necessary to scientific hydrotherapeutics.

It will be sufficient to give the analysis of the *Cardinale* spring, which is the most highly ferruginous.

CHEMICAL ANALYSIS

WATER 1000 grammes (1 litre).	SOURCE CARDINALE.
Carbonic acid, free . . . Lit.	0·225
Nitrogen with oxygen . . . „	traces
	Grammes.
Bicarbonate of magnesia	0·0761
Crenated oxide of iron	0·0980
Sulphate of lime	0·0400
„ soda	0·0060
Chloride of sodium	0·0120
„ magnesium	0·0030
Crenate of potash	0·0020
Alumina	0·0330
Traces of carbonate of ammonia.—*Girardin and Morin.*	0·2701

18

The three springs yield together 8157 gallons per day. The water has no smell, and is perfectly clear, except that of the *Reinette* spring, which is sometimes clouded. The taste is not the same at all three; it is ferruginous at the *Reinette*, and still more so at the *Royale*, but *atramentaire* (inky) at the *Cardinale*.

One of the first effects of the waters of Forges is to stimulate the renal secretion. This diuretic action explains why they have always had the reputation of curing gravel. It was as a record of the advantages which he had derived from them in this respect, that Cardinal Richelieu (who suffered from red gravel) gave his name to the spring whose waters he pre-ferred.

But the speciality of these waters lies in the treatment of affections characterised by debility of tissues, general weakness, and want of activity in organic functions. Thus, they are highly effectual in the treatment of chlorosis and anæmia, amenorrhœa, dysmenorrhæa, and passive uterine congestions. Excellent results are also produced in cases of certain intestinal disorders (gastralgia, dyspepsia, and diarrhœa), hysteria, and other nervous affections connected with impoverishment of blood. And finally, they at times justify their old reputation against sterility.

Life at Forges is very quiet, and suited rather to invalids than to persons in search of amusement.

SAINT-AMAND

(*Nord*)

SULPHUROUS MUD-BATHS—CALCAREO-SULPHATED SPRINGS

SAINT-AMAND is a small town of about 5000 inhabitants. It stands on the borders of an immense forest on the banks of the *Scarpe,* about seven and a half miles from *Valenciennes.*

This Station was well-known to the Romans, by whom it was much frequented. This is proved by the ruins of the sanatory Establishment which they erected, and by the numerous remains which have been discovered since 1740, when excavations were first commenced. Amongst these are medals bearing the heads of Julius and Augustus Cæsar, Vespasian, and Nero; a small bronze altar with reliefs illustrating the history of Romulus and Remus, and numerous other objects.

The Thermal Establishment is one of the finest in France. It is about a mile from Saint-Amand, and is capable of accommodating more than 100 patients.

The private baths and douches occupy one part of the building, whilst the other, which is of a circular shape, contains the mud-baths, divided into sixty-two separate compartments, each lined with wood, about three and a half feet square, and varying from three to six feet deep.

The springs, of which there are four, are calcareo-sulphated. The two most important are known as the *Fontaine Bouillon* and the *Pavillon-ruiné*. The water is clear and colourless, with a sulphurous smell, and a pronounced flavour of gall. It deposits in the reservoirs confervæ of *glairine*.

The water of both springs has a temperature of 21° C. (70° F.), and contains lit. 0·19 of carbonic acid. We shall give only the analysis of the *Source du Pavillon ruiné*.

CHEMICAL ANALYSIS

WATER 1000 grammes (1 litre).					SOURCE DU PAVILLON RUINÉ.
Temperature .	.	.	.	.	21° C.
Carbonic acid, free	.	.	.	Lit.	0·19
					Grammes.
Carbonate of lime	.	.	.	.	0·066
,, magnesia	.	.	.	.	0·079
Sulphate of soda	.	.	.	.	0·234
,, lime	.	.	.	.	0·870
,, magnesia	.	.	.	.	0·152
Chloride of sodium	.	.	.	.	0·018
,, magnesium	.	.	.	.	0·095
Silicic acid	.	.	.	.	0·020
Traces of organic matter and iron.—*Kuhlmann.*					1·534

The water is drunk, as well as taken in baths and douches. It is stimulating, diuretic, resolutive, tonic, and restorative. The Mud (*boue*), which constitutes the principal medication adopted at Saint-Amand, is black, with a strong sulphurous odour, and a tempera- ure of 25° C. (77° F.) Bubbles of hydrosulphuric gas continually rise to the surface, where they burst, and diffuse a sulphurous smell throughout the building.

The Mineral Mud is formed from three superposed layers of earth, the upper one consisting of clayey peat, the second of clay, and the third of silica, carbonate of lime, oxide of iron, and alumina. Through this latter, which is from seven to eight feet thick, rise numerous small springs of sulphureous water, which traverse the upper layers and convert them into mineralised mud, and this is used in the compartments already mentioned, which are employed for bathing purposes. The baths are taken every morning, the patient commencing by remaining in them for two hours, and gradually increasing their duration up to four or five hours. The compartments are so arranged that their occupants can read, write, and even breakfast, while immersed in the mud. Each patient has his own compartment, which is reserved for him during the whole course of treatment. When this is terminated the mud is removed, and replaced by some which has never been used. After each bath, the patient is carefully wrapped up, and removed on a wheeled couch to the ordinary water baths, where he washes himself.

The affections treated at Saint-Amand are numerous and varied. Amongst the chief of these may be quoted muscular and articular rheumatism, chronic gout, and paralysis; diseases of bones, such as caries, necrosis, coxalgia, and knee-joint disease; anchylosis, painful calluses, sprains, gunshot wounds;

also chronic inflammation or congestion of the uterus, with or without hypertrophy and ulceration of the neck ; and, lastly, certain forms of neuralgia, and principally sciatica. Among the diseases of the skin are those of a dry type, such as psoriasis, eczema, lepra, lichen, &c. It may be added that, as the different affections treated at Saint-Amand are almost always chronic and inveterate, the course of treatment generally lasts from five to six weeks.

Almost all the patients stay in the Establishment, so that the life is of a very quiet and domestic character. Reading, cards, conversation, and walking in the neighbourhood, are the sole amusements obtainable at this Station.

In the vicinity of Saint-Amand, students of history can visit the castles of the Princes of *Ligne* and *Croï*, and the battlefields of *Fontenoy, Bouvines,* and *Denain.*

THE West of France is almost entirely destitute of Thermal Establishments. As a matter of fact, it has but two, that of *Bagnoles-de-l'Orne* and that of *Château-Gonthier*.

BAGNOLES-DE-L'ORNE

(*Orne*)

MIXED SULPHURETTED SPRING (FEEBLY MINERALISED)—
FERRUGINOUS SPRINGS.

BAGNOLES is a charming village of 606 inhabitants, situated in a valley which, from its formation and picturesque aspect, has been called "the Norman Switzerland." Bagnoles is about fifteen and a half miles from *Alençon*.

The waters, which have been in use for nearly two centuries, were discovered, if the legend may be believed, under very curious circumstances. An old horse, a confirmed roarer, abandoned by its owner, wandered into the valley. Passing through it, some months afterwards, its master saw and recognised his old servant. The horse went up to him; and to his astonishment he perceived it was fat, and had a smooth and glossy coat. Curious to find out the cause of so astonishing a cure, the man watched the movements of the animal, and observed that it went to drink at the foot of some black rocks. Dipping his hand into the water, he discovered it to be a hot spring, clear and abundant.

The springs are three in number, one sulphureous, discharging about 24,250 gallons per day, at a temperature of 27·5° C. (81·5° F.), and two ferruginous springs of minor importance. The Establishment in which they are utilised consists of forty private bath-rooms and two large swimming-baths (the latter supplied with continually running water), douches, shower baths, &c., and a *buvette*. Bathing and drinking the water are almost invariably combined in the treatment.

The chemical analysis of these springs has not been made in a very accurate manner; that prepared by Vauquelin in 1812—which mentions salts of soda and magnesia, and sulphate and chlorohydrate of lime—does not give the quantities which were present. As to the physical properties, the water is clear, colourless, oily to the touch, and has a pleasant taste, with a very slight sulphurous smell in the glass, but a stronger one in the bath. The affections for which it is prescribed with most success are hysteria, chorea, palpitations of the heart, and spasms. Good results have been obtained in cases of stiffness of the joints, the commencement of anchylosis, and certain degenerations of the osseous tissue, also in gouty and rheumatic affections. Bagnoles water has also a well merited reputation for the cure of certain skin diseases,—such as eczema, psoriasis, prurigo, &c. Were this Station developed by a company with sufficient means, it might attain to an importance which would be justified by the reputed value of its waters.

The vicinity of Bagnoles is charming, and offers several interesting objects for walks. Amongst these are the Chapel of *Saint-Horter;* the ruins of the *Château d'Ambrières,* which belonged to William the Conqueror; the old prison of *Bon-Vouloir; Domfront,* with its crenelated towers; the *Roc-du-Chien,* an immense mass of granite; and the old châteaux of *Lassey, Couterne, Carrouges,* &c.

CHÂTEAU-GONTHIER

(*Mayenne*)

COLD MIXED FERRUGINOUS SPRING

CHÂTEAU-GONTHIER is a little town in the Department of Mayenne, which possesses a cold ferruginous spring called the *Source Rouillée,* containing bicarbonate of lime and sulphate of magnesia. This water supplies a somewhat important and well kept hydro-pathic Thermal Establishment, which comprises a number of baths, several varieties of douches, and vapour baths.

The waters of Château-Gonthier are suited to persons of slow and difficult digestion, and to females suffering from chlorosis, amenorrhœa, and dysmenor-rhœa. It is said that they have been prescribed with success in cases of catarrhal affections of the bladder.

PYRENEAN GROUP

ARCACHON

(Gironde)

THIS is a small town of 3600 inhabitants, situated between the bay of *Arcachon* and the vast forest of pine trees to which it owes its name. (In the *patois* of the country, *arcanson* or *arcasson* has the meaning of resin.) It is easily reached by train from Bordeaux, the journey occupying about seventy-five minutes. A favourite bathing place in summer, when it is visited yearly by over 100,000 persons,—it is, in the cold season, a winter Station of considerable importance. The climate of Arcachon is equable, and very similar to that of the most favoured Mediterranean health-resorts. The mean temperature is for the whole year 15° C., or 59° F. (slightly inferior to that of Nice); in winter it is 46·4° F. on the sea shore, and 50° F. in the forest. The winter residences—forming what is called the winter town—have been built in the forest, in order to shelter them from the north winds, to which the shore and town are exposed. No better spot could possibly have been selected than this magnificent forest of pine trees, nearly 9000 acres in extent, forming an impenetrable screen against all strong winds, and affording an endless variety of pleasant walks along its innumerable avenues and lanes.

The accommodation to be found in the *Ville d'Hiver* is generally good. Besides several first-class hotels, a large number of villas and elegant and comfortable *châlets* are to be had at prices which are, as a rule, moderate, and vary according to the more or less sheltered position of the house. These are let by the month or season, and it is advisable to make all arrangements before entering, in order to avoid trouble or misunderstandings afterwards. The water supplied to the town comes from an Artesian well 370 feet deep, and is very pure and of good quality.

The climate, being sedative, is suitable for patients suffering from chest affections, winter cough, catarrh, chronic bronchitis, and asthma.

Arcachon is celebrated for its oyster-beds, which are one of the most interesting sights of the place; it is said that from eight to ten millions of oysters are sold annually. Excursions in the environs can be made both by sea and land, and patients are recommended to ride or drive in the forest as much as possible, in order to breathe the healthy and invigorating air of the pine-wood. The town is well supplied with good horses and carriages, which can be hired at very moderate prices.

DAX

(*Landes*)

DAX stands on the left bank of the river *Adour;* it is a town of 10,000 inhabitants, on the main line from Paris to Madrid, the journey from Bordeaux being accomplished in three hours and a half by express train. Dax is a very old town, and has been known for a considerable time for its abundant sulphureous springs, the temperature of which ranges from 87·8° to 141·8° F. It is also renowned for its mud-baths, and is now growing in favour as a winter health-resort, the climate and situation being eminently favourable for the cure of many affections.

The air is milder than that of Pau, the mean winter temperature being 8·3° C., or 46·6° F. The hot springs, which are very numerous, no doubt contribute to warm the atmosphere, while the surrounding forest of pine trees protects the town against all strong winds. The rainfall is moderate, and snow is very rarely seen.

Chronic rheumatism in all its forms and complications—whether neuralgic, muscular, or articular—is treated at Dax with the greatest success; scrofula and rickets are also much benefited. One of the great advantages of this Thermal Station is that patients who are required to follow a winter hydro-mineral treatment, and whose power of locomotion is impaired, can reside at the *Établissement des Thermes,* which is specially adapted to all the requirements of invalids, and is under the immediate supervision and direction of eminent medical men, who have made the class of diseases treated there their special study, and have largely contributed to the success and reputation of this Establishment. The whole of the building (in which a uniform temperature is maintained throughout), is heated by pipes, through which the water of the sulphur springs is made to circulate, by means of a very ingenious system. Large and well ventilated galleries in which patients can take walking exercise, or be driven in bath chairs at all times, make them quite independent of the changes in the outer temperature. Winter residence at Dax is also particularly well suited for persons suffering from pulmonary diseases.

PAU

(Basses-Pyrénées)

PAU is the most renowned of all the Pyrenean winter health-resorts. It is a fine town of nearly 29,000 inhabitants, in an admirable situation, about 700 feet above the level of the sea, seventy-five miles from the ocean, and fifteen miles from the Pyrenees. The three principal streets of Pau run in a parallel direction, towards the terrace of the Château overlooking the valley and *gave* or river, from which there is a splendid view of the chain of the Pyrenean mountains, the highest peaks of which are covered with snow during the greater part of the year.

Owing to its exceptional climate, Pau is every winter selected as a residence by a great number of invalids. The climatic advantages of this town are the almost total absence of strong winds, the dryness of the atmosphere, and its great equability of temperature, which in winter is on an average $43 \cdot 2°$ F. Although the rainfall is considerable, more especially in the autumn and spring, the rain dries up very quickly, owing to the gravelly soil; and in a very short time after a heavy shower the streets are dry enough to allow patients to go out. Frosty weather is common in winter, and snow occasionally falls at Pau; the nights also are cold. The most prevalent winds blow from the north and west. Those from the north-west indicate dry, cold weather, while westerly winds are the forerunners of wet weather or heavy showers. Easterly winds are very rare. Pau is well drained, and its sanitary condition quite satisfactory. It is, no doubt, for this reason that the town is remarkably free from epidemics of all kinds.

The climate is essentially sedative, and therefore very favourable for the treatment of phthisis in its early stages, and also for persons suffering from nervous and vascular irritation,—it is likewise beneficial in irritations of the mucous membrane of the air passages, or of the alimentary canal; but it is unsuitable, as being too relaxing, in chronic bronchial catarrh with profuse expectoration, in obstinate uterine or intestinal discharges, and in all disorders attended with congestion of the venous system and diminished nervous energy.

The season at Pau commences in September, and ends in May or June. A large number of first-class hotels, furnished apartments, and villas, both in and out of the town, afford ample and very good accommodation.

Pau was formerly the capital of Béarn, and is now the chief town of the Department of Basses-Pyrénées. It possesses several fine buildings, the most important being the *Château*, in one of the rooms of which Henry IV was born in the year 1535. The many interesting relics, fine old tapestries, and specimens of art in the Château are well worth more than one visit. The churches of *Saint-Jacques*, *Saint-Martin*, and *Saint-Louis*, the museum, the Palace of Justice, and the *Hôtel-Gassion*—now transformed into a casino and winter garden—are amongst the other edifices worthy of notice. There are in Pau several English churches, and an excellent English club. Foremost among the attractions of Pau must be mentioned its admirable promenades, such as the *Place Royale*, the gardens of the Château, the park, with its fountain and gymnasium, and the *Bois Louis*, near the railway station.

BIARRITZ

(Basses-Pyrénées)

BIARRITZ, well known as a fashionable sea-bathing place, is a town of about 4000 inhabitants, on the main line from Paris to Madrid; it is built on a high cliff, some parts of which are more than 120 feet above the level of the sea. It has lately become a favourite winter residence, much frequented by the English, who, during the cold season, are in the majority. The air is generally mild; strong winds from the north and east are very rare, and although snow occasionally falls, it never remains on the ground. The climate is tonic, but not exciting, and owes its equability to the Gulf Stream, the powerful influence of which is felt in the greater part of the Bay of Biscay. As the atmosphere of Biarritz contains a moderate quantity of moisture without being damp, it is an excellent place for patients to spend the winter in, after having undergone a hydromineral treatment at one of the numerous Pyrenéan watering places.

Hotel accommodation at Biarritz is very good, and in nearly every house there are comfortable furnished apartments to be let by the month, or for the whole season. There is an English church at Biarritz, and a British vice-consul resides in the town.

Numerous excursions can be made in the neighbourhood of Biarritz which are particularly interesting to

the English visitor, the whole of this part of France
having been included in the Peninsular war. Very
few changes have taken place since that period, as
testified by the descriptions to be found in that
excellent little book "The Subaltern," which is even
now as good a guide to the Basque country as one
could be desired.

A small line of railway has been constructed to
connect Biarritz with *Bayonne*, a very picturesque
old town on the rivers Adour and Nive, which should
be visited, as it is of historical interest. Trains run
every half hour.

SAINT-JEAN-DE-LUZ

(*Basses-Pyrénées*)

Patients who like a more quiet life than that of
Biarritz may go to *Saint-Jean-de-Luz* (twenty minutes
by train, one hour by road from Biarritz), a pleasant
old Basque town, in a very agreeable situation at the
end of a semicircular bay. Saint-Jean-de-Luz has
precisely the same climate as Biarritz, and possibly
may be a little more sheltered against east and north
winds. There are several good hotels, and a number
of furnished apartments and villas to be let at
moderate prices. A new English church was opened
on Christmas Day, 1882. In this old town, Louis
XIV of France was married to the Infanta Maria
Teresa of Spain in 1660. The houses in which they

resided during the marriage festivities, as well as the curious church in which they were married, are very good specimens of old Basque architecture.

The same excursions can be made from Saint-Jean-de-Luz as from Biarritz, such as the ascent of *La Rhune*, a mountain 3000 feet high, from which there is a magnificent view over the range of the Pyrenees, —or a trip to Spain, the frontier being only at twenty minutes' distance by rail. The nearest and most interesting Spanish towns within easy reach by rail or road, are *Fuenterrabia* (Fr. *Fontarabie*) and *San Sebastian*.

It is not advisable to go out in boats, either from Biarritz or Saint-Jean-de-Luz, as the coast is rocky and dangerous, and the sea generally rough.

AMÉLIE-LES-BAINS

(*Pyrénées-Orientales*)

AMÉLIE-LES-BAINS is situated in the valley of *Mondony*, in the Pyrenees, about 550 feet above the level of the sea, and within eighteen miles of the Mediterranean. Owing to its climatic advantages, Amélie-les-Bains, which is much frequented in summer on account of its sulphur baths, is fast becoming a winter health resort of more than average importance. Being the southernmost Pyrenean Thermal Station, it has a very mild and equable winter climate. In the autumn and winter months, the variations are slight during the day-

time, the rainfall is moderate, and there are no fogs.
The surrounding mountains protect the village from
the north winds, and the *mistral* is very seldom felt.
The mean temperature from November to January is
9° C., or 48·2° F., and from February to April, 11·6° C.,
or 52·7° F.   The number of rainy days for the corresponding periods is twelve and eleven respectively,
and the average number of fine days 210 yearly.

The medical day (that is, the time during which
patients may take outdoor exercise) commences at 10
o'clock in the morning, and ends one hour before sunset, and no invalid should venture out in the evening.
The climate of Amélie-les-Bains being moderately sedative, patients who find Pau too relaxing and the Riviera
too exciting will do well to try this Winter Station.
There are several good hotels and *Établissements* at
Amélie-les-Bains, and a number of furnished *châlets ;*
the best being situated on the promenade called *La
Petite Provence.*

It will thus be seen that Amélie-les-Bains differs
from most Thermal Stations in being independent of
the seasons ; so that patients who, from whatever cause,
could not follow a hydromineral treatment in the
summer, are able to find in this privileged spot the
combined advantages of a sulphurous water cure and
of a winter spent in a mild climate.   (See Amélie-les-
Bains, page 155.)

LE VERNET

(Pyrénées-Orientales)

LE VERNET, although in a much higher situation than Amélie-les-Bains (about 2000 feet above sea-level), has, nevertheless, an equally mild climate. It is certainly not to be compared to the Mediterranean health-resorts, but it is unrivalled as a Thermal Station, where, at all times of the year, patients can take the waters, and follow a hydromineral treatment. This is a great advantage to those who are unable to leave their homes in time for the usual summer season, and would thus be precluded from following the prescriptions of their medical adviser for more than six months, were it not for such places as Le Vernet.

In its stimulating effects, the climate resembles that of Amélie-les-Bains ; but the difference of altitude is of great importance. Some patients, even when suffering from emphysema or dyspnœa, can bear the air of the mountains better than that of the valleys or the sea shores ; while others, on the contrary, are only well under a heavier atmospheric pressure. Such idiosyncrasies will in each case determine the choice between the two stations just described. (See Le Vernet, page 153.)

MEDITERRANEAN GROUP

HYÈRES

(*Var*)

THE little town of Hyères, or as it is also called—
Hyères-les-Palmiers—stands on the southern slope of
a hill, well protected against the north winds, and faces
the sea, from which it is about three miles distant.
The new town lies from west to east, on the road from
Toulon to *Saint-Tropez*, and overlooks a valley extending from *Cape Bénat* to the *Colline-des-Oiseaux*,
the sea shore forming the southern boundary of the
valley.   Hyères is the southernmost and sunniest
spot on the coast of Provence, and is well sheltered
from the *mistral* by a chain of hills stretching from
east to west.   The climate is milder than that of
Nice; it is warm, dry, and tonic.   Being situated
at a short distance inland, Hyères is sufficiently
free from the stimulating effect of the immediate
vicinity of the sea, whilst it is not beyond the salutary
influence of the atmospheric movement which tempers
the heat and gives to the air greater purity.   The
average temperature in the winter season is 12·3°

C. or about 54° F. There is a difference of opinion
as to the prevalency of winds, and especially of the
mistral, some being of opinion that Hyères enjoys a
complete immunity from it, whilst others assert that it
is often felt during winter. In the absence of reliable
scientific observations, it is difficult to speak authorita-
tively on the subject, but it can be safely assumed
that there is much exaggeration on both sides. It is
certain that the *mistral* is occasionally felt at Hyères,
as well as in all Mediterranean Winter Stations, and
perhaps, on an average, rather more than at Cannes
or Nice. The amount of rainfall is low, fogs are
unknown, and snow is very seldom seen.

Owing to its climatic advantages, and its quiet life,
free from excitement and noise, Hyères particularly
recommends itself as a Winter Station to persons of
delicate constitution, or to those whose health, impaired
by illness, requires the benefit of a warm climate
where they can without danger enjoy the bright sunshine, inhale the pure air, and take moderate exercise.
Patients suffering from phthisis, rheumatism, and
debility, constitutional or acquired, will derive the
greatest benefit from the climate of Hyères. As in
all winter health resorts, patients must take precautions, when taking out-door exercise, against the
changes in the temperature experienced at sunset, or
when passing from the sun into the shade. They must
wear warm woollen clothing, but light in weight, and
walk at a moderate pace. The medical day practically
begins at 10 a.m. and ends at 3.30 p.m. It is unwise

to sit in the shade in the open air. Invalids should never be out after sunset.

As a rule, patients will do well to avoid remaining in a northern country until the cold weather sets in ; and it is important not to hasten home too soon, but to wait until the warm weather of May has nearly equalised the temperature of the place they leave and that of the country to which they are returning. It is needless to insist on the prejudicial effect to persons in delicate health of exposing themselves to the cold and unsettled weather of an English April, immediately after leaving a dry and warm climate such as that of Hyères.

Every season, there are about one thousand English families wintering at Hyères, where excellent and ample accommodation is to be found in the new town, chiefly composed of hotels, boarding-houses (*pensions*), and private residences. Numerous excursions can be made in the neighbourhood, the most pleasant being within walking distance. A fashionable promenade is the *Jardin D'acclimatation.* The harbour is very fine, and communicates with the sea by five different channels formed by the Îles d'Hyères, which are a favourite object for sea excursions. Boats can easily be hired to go to the islands, but there is as yet no regular communication between them and the town.

CANNES

(*Alpes-Maritimes*)

CANNES is a most admirably situated health resort, and it is a matter of considerable surprise that for so many years it was an unknown, insignificant little town of some 4000 inhabitants. Within the last thirty years it has acquired great importance, and it is now one of the most favourite Winter Stations on the Mediterranean. It is annually visited by thousands of patients, for whom ample hotel and private accommodation has been provided by the inhabitants, who are keenly alive to the necessity of keeping up the reputation of this pleasant and health-giving locality.

Cannes stands in a remarkably favourable situation, at the end of a small horse-shoe shaped bay, along which it extends for a distance of nearly four miles,— and is well protected against the north and north-west (*mistral*) winds by the hills of *Grasse* and *l'Estérel*, whilst the slopes of *Le Cannet* and the promontory of *La Croisette* form a most effective barrier against the easterly winds. The opening of the bay is nearly due south, and admits the warm and balmy breezes of Africa. The climate is delightful; it is very mild and equable, and the average temperature during the season extending from the beginning of November to the end of April, is 12° C. (53·6° F.), with variations more moderate than in any other Mediterranean Station.

From a medical point of view, Cannes is divided into four zones or regions; namely, the sea zone, the intermediate zone, the hills, and *Le Cannet*, a small village about two miles from Cannes. The first, or sea zone, is the most suitable region for the air cure. Every day the bay is covered with a flotilla of small boats conveying the patients to the Islands of Lérins, where they spend the day under the shade of the pine trees, and inhale the vivifying sea air. The return takes place in time for them to be in their dwellings one hour before sunset, and this is a most important point, which is constantly to be borne in mind.

Under the combined influence of a sunny climate, of the purest air, and of the sea breeze, the digestive organs are strengthened, the circulation becomes more active, and the nerves stronger. The success of this mode of treatment is almost certain, provided the patient can bear the stimulating action of the sea air. The same climatic influence is to be found all over this region, but in a diminished form. It is less active on the coast, and still less in the second or intermediate zone; it is felt again upon the hills, and practically disappears at Le Cannet and in its neighbourhood. The air in those regions is more calm and mild, and patients suffering from extreme debility should be sent to the zones farthest from the sea. The choice of a residence is a matter of the utmost importance, and the success of the cure depends in a great measure on the careful selection of a suitable place of abode. This is a matter which patients should on no account take upon themselves to decide, but should reserve for the consideration of their local medical adviser.

The autumn and winter months are the best and most favourable for the *air cure*, and patients will do well to take up their winter quarters as early as possible. Unfortunately, a great number come too late, and in certain cases the sudden change of temperature may be injurious.

METEOROLOGICAL AVERAGES, ETC.

Months.	Mean height of Barometer.	Mean Temperature, *in the shade.*
November to January .	760·10 mm. (about 30 in.)	10·33° C. (about 50·6° F.)
February to April .	759·07 mm. about 29·9 in.	11·56° C. (about 52° F.

Northerly winds are frequent, but are not dangerous, Cannes being well protected on the north side. They generally indicate the approach of fine weather. North-east winds, when strong, are cold and disagreeable, and during their prevalence it is advisable not to go out.

South-east winds, without being injurious, are the forerunners of rain.

South winds (sirocco) are very rare, and in winter are rather beneficial to patients suffering from chlorosis and anæmia, being saturated with saline vapours, but do not suit consumptive patients.

The *mistral* generally blows from the north-west,

and seldom before the month of March. Patients who during its prevalence have to remain indoors, may increase the moisture of the atmosphere of their rooms by having in their apartment a basin containing boiling water and a few drops of essence of eucalyptus, and allowing the vapour to escape.

Food and clothing, duration and direction of walks, are all matters worthy of attention,—considering the different mode of living, the new surroundings, and the rapid transition from an English to a southern climate. Woollen clothing will be found preferable to any other, and when walking an overcoat or shawl should be carried, to be worn when sitting or standing, or passing from the sun into the shade. The use of a parasol to protect the head against the rays of the sun is strongly to be recommended.

It is well known that in warm climates lighter food is required than in cold or damp countries. It is, therefore, advisable to adopt a diet as appropriate as possible to the climate, whilst in conformity with the patient's usual mode of living.

As to walks, they should be as frequent as practicable, and the state of the weather and direction of the wind must be taken into account. In winter, ten o'clock in the morning will be found early enough for a first walk, and no patient should venture into the open air after sunset.

Excellent sea bathing can be obtained on the shore, the temperature of the water being, on an average, 6° C. (10·8° F.) higher than in the Atlantic. Bathing in the sea can be permitted to a certain class of patients as early as March, but the real bathing season is from the 1st of April to the middle of June.

The neighbourhood of Cannes abounds in lovely walks, and very pleasant excursions can be made by sea to the Islands of Lérins, which are easily reached by boats running at regular intervals. But the chief characteristic of Cannes is that (owing to the absence of theatres and other places of amusement), it is absolutely free from noise or excitement, and on that account particularly well suited for invalids and persons in delicate health, who are certain to find there the calm and quiet which is so necessary to them, and sometimes wanting in other Winter Stations.

GRASSE

(*Alpes-Maritimes*)

GRASSE (13,000 inhabitants) is delightfully situated on the southern slope of the *Rocavignon*, nearly 1000 feet above the level of the sea; it is chiefly known as the most important town in France for the production of floral essences of all kinds used by distillers and perfumers, and the wonderful gardens whence the flowers come which supply this particular industry.

The climate of Grasse is very mild, although in winter the temperature is rather lower than that of Cannes or Nice, and is of great value for patients who are unable to bear the over-exciting influence of the sea. The mean temperature is, for the whole year, 60° F., and for the winter 46·4° F. Being encircled by high hills, Grasse is well protected against the *mistral* and easterly winds, and the nights are not so cold as in the valley below.

In spite of its mild and equable climate and admirable situation, Grasse has hitherto been neglected by visitors, on account of its want of proper accomodation. This defect has now been remedied, a very good hotel has been erected, and numerous villas are being built, which will afford sufficient accommodation for the patients who now begin to winter in this quiet and pleasant town.

Grasse possesses several old and interesting monuments, and a fine library. It is connected with Cannes and the main line by a branch railway.

NICE

(*Alpes-Maritimes*)

NICE unquestionably enjoys a greater reputation than any other Winter Station, and, it must be acknowledged, fully deserves the high renown it has acquired for itself ever since the days of Nero. It is very picturesquely situated at the eastern extremity of a small bay, termed *Baie des Anges*, and is divided into two unequal parts by the *Paillon*, a small stream,—or rather bed of a stream, nearly always dry, and (it must be confessed) a deplorable eyesore.

The old town on the left bank is chiefly inhabited by the native population, and the new town, on the right bank, is in winter frequented by over 50,000 foreign residents, coming from all parts of the world. This large influx of visitors is due to the many advantages which this privileged city possesses in a degree unequalled by any other health resort.

Looking fully south, in the direction of the sea, Nice is admirably protected against the north, east, and west winds by a double semi-circular range of hills, extending from the river Var to the Bay of Ville-franche.

The climate is dry, the rainfall insignificant, and the atmospheric variations very moderate and regular. Considering that in a climatic health-resort equability of temperature is of the highest importance, there is no other Winter Station more favorably situated.

The average temperature during the invalid's season (from November to April) is 10·65° C. or 51° F. Snow falls very rarely at Nice, and fogs are unknown. Winds, on the other hand, are sometimes troublesome, especially at night; and in the daytime they often raise clouds of dust. The most windy months are March and April; and although Nice is comparatively free from injurious winds, the *mistral* is occasionally felt during these two months, and patients will do well to remain indoors while it lasts. Another characteristic feature of the climate of Nice, but one which it shares in common with all Winter Stations, is the great difference in the temperature experienced when passing from the sun into the shade; this is due mainly to the dryness of the atmosphere and the purity of the sky, which give to the solar radiation an unparalleled intensity.

Medically speaking, each quarter of Nice may be said to have its particular climate, for there are in reality three distinct climatic zones,—namely, the marine zone, the valley, and the hills, which are respectively stimulating, sedative, and tonic. The marine zone is particularly suitable for children and aged persons; the valley is the most favourable position for invalids in general; and the hills will be found the best situation as a residence for patients whose nervous system is very excitable, and who can not bear the stimulating influence of the sea air. It is therefore a matter of paramount importance to persons intending to winter at Nice, to select a suitable place of abode, as the success of the cure depends in a great measure on the judicious choice of a residence. Patients have frequently been known to return home no better, or even worse, than when they first went to Nice, simply because they took upon themselves to choose the part of the town in which they would spend the winter season, instead of consulting a resident physician. Invalids wintering in the south should remember that it is as injudicious for them to solve this question without competent advice, as it would be to prescribe for themselves.

There is no difficulty in finding excellent accommodation in any of these three zones, as Nice is well provided with a large number of hotels, pensions, and furnished villas.

It is indispensable that patients should arrive early in the winter, the too sudden change from one climate to another being in some cases very injurious, and in all cases objectionable. Dress and food are matters of importance, and winter clothing should not be left behind, as it may occasionally be required. Invalids are recommended to wear warm—but not *heavy*—clothes, and to carry an overcoat or wrap, to be worn when in the shade or driving. Broad-brimmed hats and parasols are indispensable, as the head should never be exposed to the burning rays of the sun; and coloured glasses will be found very useful, as a protection to the eyes against the full blaze of sunshine. As to diet, it is well to remember that lighter food is required in warm climates than in cold countries. Strong wines, spirits, and rich dishes are unnecessary, and even injurious. At the same time, a radical change from the usual mode of living should not be attempted, save in some exceptional cases, and a judiciously modified diet will be found sufficiently appropriate to the climate.

Out-door exercise should be taken regularly, and there are precautions to be strictly observed by all invalids, which we will briefly enumerate. Patients must not be out of doors before ten o'clock in the morning, and ought to be indoors an hour before

20

sunset. It is advisable to walk at a moderate and steady pace, and not to indulge in violent exercise as some people do,—in order to avoid perspiration, and the consequent risk of a chill, if exposed to a gust of cold wind. When driving, shady roads should be avoided, and an ample supply of rugs and shawls should always be taken.

The dining and reading rooms in hotels and other similar establishments, being generally on the north side of the house, are as a rule chilly; and patients have caught more colds in them than would be imagined, especially in the middle of the day, when coming in after a walk in the sun and sitting in a cold draughty apartment. It is therefore recommended to wear warm clothing when at meals; and finally, invalids should avoid going out of the house after sunset. In the evening, a small wood-fire should be lighted in the room; it makes the apartment more cheerful and healthy, by renewing the supply of fresh air and increasing the ventilation.

The curative properties of the climate of Nice are mainly due to the purity and dryness of the atmosphere, the slightness of the variations in the temperature, its immunity from severe cold, and the greater proportion of sunshine it affords to invalids, who can enjoy out-door exercise during the whole of the winter. For these reasons, Nice as a winter resort is most suitable for consumptive patients, and persons suffering from from catarrh, asthma, anæmia, rheumatic gout, and scrofula.

METEOROLOGICAL AVERAGES, ETC.

Months.	Mean height of Barometer.	Mean Temperature.
November to January .	761·80 mm. (about 29·99 in.)	10° C. (about 50° F.)
February to April	760·80 mm. (about 29·95 in.)	11·75° C. (about 53° F.)

Annual rainfall, 811 mm. (about 31·62 inches.)

There are on an average, during the year, 60 rainy days, 180 days of brilliant sunshine, and 125 cloudy days.

The north, south, and south-east winds are moist; while those of the west and north-west (*mistral*) are very dry. The latter is on an average felt about ten times during the year. November, December, and January are the least windy months; February, March, and April, being those during which winds are most prevalent.

Apart from its many advantages from a medical point of view, Nice has all the attractions and comforts of a great city. It is every day growing finer and larger, its numerous promenades (the best known of which are the *Jardin Public*, the far famed *Promenade des Anglais* and the Pier), its theatres (three in number), clubs, reading rooms, public library, and museum, have secured for this favourite Mediterranean city the patronage of the most refined society, to whom it affords a variety of amusements, combined with a delightful climate. It is quite true that invalids are more in want of quiet and rest than of noisy entertainments; at the same time it cannot be denied that the various and animated scenes to be met with when out in the open air, and the numerous bands of music playing in the public gardens, have a certain beneficial effect in cheering patients of a desponding disposition, and drawing their thoughts from their ailments, over which many may have a tendency to ponder. For them, as a matter of course, evening entertainments are out of the question; but there is no reason why a health-resort should not be made as attractive as possible, and in this respect Nice leaves nothing to be desired.

Nice has in its immediate neighbourhood several charmingly situated districts, where the utmost quiet can be enjoyed while remaining within walking distance of the centre of the town, and of all its conveniences and advantages.

MENTON

(Alpes-Maritimes)

MENTON (Italian, *Mentone*) is a charming little town of 8000 inhabitants, delightfully situated on a promontory which divides into two equal portions the semicircular bay of the same name. There are two distinct quarters, the old town and the modern town, the latter being built around the former, and nearer the sea shore.

The climate of Menton is rather warmer than that of
Nice, and closely resembles the climate of Cannes. The
town is well sheltered from the north and east winds
by the range of hills which surround it on three sides,
the only opening being on the south. The average
temperature during the season (1st of November to 1st
of May) is 12° C., 53·6° F. The temperature is very
equable, the variations very slight, and the rainfall
inferior to that of Cannes. The annual average number
of rainy days is eighty, and of clear sunny days over
two hundred; whilst the number of cloudy days is
from sixty to eighty. Like the other Mediterranean
health-resorts, Menton possesses several zones, each
having different curative properties. Here also, it is
recommended that patients, before taking up their
abode, should consult a local physician as to the part
of the town in which they are to select their residence.
The climate—which is sedative without being relaxing
—is particularly suitable to consumptives, and to
patients suffering from bronchial catarrh, asthma,
rheumatism, gout, and diabetes. One of the chief
advantages of this health-resort is that the night
temperature seldom falls very low.

Menton affords ample and very good accommodation,
both in hotels and villas, which are numerous and
generally very comfortable. Prices are high, but not
extravagant; while as to sanitary arrangements, Menton can compare favourably with the other Winter
Stations in this region.

Very few places are more advantageously situated

with regard to excursions both by land and sea. The environs are most interesting, and the varied and almost tropical vegetation is truly marvellous. When making sea excursions, patients should avoid going more than two miles from the shore; as, beyond that distance, the bay is no longer protected against the north winds, and has sometimes a much lower temperature, to which it is dangerous for persons in delicate health to expose themselves.

CONCLUSION

ADVICE TO INVALIDS GOING TO THE MINERAL WATERS,
OR PERSONS WINTERING ABROAD.

Before concluding our task, we desire to offer a few hints, based on experience, which, we trust, will materially assist patients in making their journey as pleasant as possible,—and, when arrived, enable them, to derive the greatest benefit from their Hydromineral treatment, or a more or less prolonged stay at some Winter Station.

The first point on which we ought to insist is :— *never to go to any Mineral-Water Station without first consulting a physician.* Whatever friends or acquaintances may say, however much they may have been benefited by taking the waters at such or such a place, turn a deaf ear to their well-intentioned but very dangerous advice. Mineral waters are just as much a remedy as any medicine compounded by chemists; and their properties are as different as those of various prescriptions, although their composition appears to be the same, or nearly so. It is the *nearly so* which makes the great difference, to be fully realized only by those who have made a special study of this important matter.

When once the physician's advice has been obtained, and the Mineral Station decided upon, preparations should be made in view of the approaching journey. As these will occupy a few days, especially in the case of ladies, this interval will be best turned to account by writing to · one or two hotels at the place selected, stating what apartments are required, and asking for terms; for it is important to make arrangements beforehand, nothing being more troublesome than to arrive in an unknown locality, and to be obliged to put up for a few days at some third-rate hotel, for want of accommodation at the better establishments.

The terms in hotels at Watering Stations range from twelve to twenty-five francs per diem, including board. That comprises a light breakfast (tea or coffee) in the morning, a *déjeuner à la fourchette* at eleven or twelve, and dinner at six or seven o'clock. The difference in prices depends solely on the size, comfort, and situation of the rooms; the lower floors being the most expensive. This point should be settled on arrival at the hotel, in order to avoid possible misunderstanding afterwards.

Night travelling is to be strictly avoided. The tidal trains from London to Paris (*via* Folkestone and Boulogne) are the most convenient for several reasons. The hours of departure are generally more suitable to persons in delicate health, and the time occupied in the journey is shorter than by other routes. Another advantage is that the steamers are larger and more comfortable than on the other lines ; they have also very convenient general deck saloons, and a greater number of private cabins. (It might, perhaps, be hinted here—from personal experience—that invalids will generally be allowed a reserved compartment to Folkestone, by applying the day before their departure to the courteous station master at Charing Cross.) Finally, the tedious railway journey from Boulogne to Paris is less by one hour than that from Calais.

When arrived in Paris, it is a good plan to send a telegram to the hotel at the Watering Station, to inform the manager of one's arrival, prepaying the answer in order to ensure a prompt reply. This being received, the journey can be safely resumed, always taking care to travel by daytime. In hot weather, however, and especially when going to the Auvergne or Pyrenean springs, the journey being a long one and the heat sometimes overpowering, it is better to leave Paris by the night express or *rapide* trains. The fatigue will be more than compensated for by the comfort of travelling in comparative coolness, and freedom from dust and the scorching rays of the sun.

The first thing to be done when arrived at the Springs, is to call upon the physician to whom an introduction has been obtained at home, in order to take his advice as to the mode of treatment to be adopted.

Always be guided by his advice, and do not listen to suggestions made by persons you may meet at the hotel or elsewhere. Do not be tempted to take an inordinate quantity of glasses of water, with the idea that it will hasten your cure ; for in many cases it might be not only useless but even dangerous.

It may here be mentioned—in order to save disappointment—that Mineral Water Stations are often to be found in small towns or villages where the resources are, as a matter of course, very limited. Some springs, indeed, have hardly any buildings near, except the various Hotels and Establishments which are open only during the bathing season, and are absolutely deserted during the other months of the year. It would, therefore, be unreasonable to expect to find in such places a number of articles which are so essential to our comfort and so common in our dwellings that we think we can find them everywhere.

Among other things, patients will do well to take with them the following objects, which will be found useful :

A thermometer with English and French divisions (Fahrenheit and Centigrade).

A small aneroid barometer.

A pocket compass for excursions, and for ascertaining the position of rooms, direction of wind, &c.

A spirit-lamp and apparatus for boiling water, are exceedingly convenient, and almost indispensable for invalids.

The diet is a matter of considerable importance, and has to be carefully regulated according to the course of treatment adopted, and with due regard to the age and usual mode of living of patients, who should not be subjected too suddenly to a radical change in their habits. Unless prevented by their state of health, patients should take their meals at the *table d'hôte*, where they will be better attended to, and will meet pleasant company.

English people, as a rule, entertain a sort of distrust of acquaintances met with at watering-places, and this feeling it is not our object to discuss. It should, however, be distinctly understood that in France, by a tacit understanding, acquaintanceships so made do not imply any further intercourse, or even simple acknowledgment, when meeting elsewhere. It is not necessary to exchange cards; and people who meet at the Establishment every day, and to all appearances are on almost intimate terms, often do not know each other's names, and will not think of addressing one another when in town. The obvious advantage of this custom is that it enables people who are thrown together in a strange place to enjoy social intercourse, whilst preserving complete liberty.

The usual stay at a Mineral Station is three weeks, that is, twenty-one days of actual treatment. When this is over, do not, however, hasten back too quickly. The medicinal effects of the waters continue for some time afterwards; and, if possible, a week's rest at some quiet place on the way home will be most advantageous to all patients, and in some cases is absolutely indispensable.

Another important recommendation that we would make is—not to rush into business or pleasure immediately on arriving in town, but to resume your usual mode of living gradually. In many cases it will be necessary, in order to complete the cure, to continue at home the use of the mineral waters, but your physician at the Station will give you the required instructions.

We will now add a few words on the precautions to be taken by persons in delicate health going to winter abroad.

What we have already said with regard to the mode of travelling is equally applicable to them. As a rule, they should leave England not later than the 1st November, and as early before that date as convenient. They should bear in mind not to return too quickly to a northern climate, in order to avoid the sudden change of temperature, especially if they leave *early* in the spring : this has often proved dangerous, and the proper time to come home after spending the winter in the south is the middle or latter end of May.

As to the hotel or place of dwelling to be selected at Wintering Stations, this is a matter which (for reasons we have already explained) cannot be left to the patient, who must consult a resident physician. The best course to adopt is to engage rooms at one of the hotels for a few days, until a suitable place of residence has been decided upon.

Our task is now completed : and we have only to bid farewell to our readers, in the hope that those who have consulted this book—whether on their own behalf or as a guide to them in advising others—have not failed to find in it any needed information.

APPENDIX

LEGISLATION RELATING TO MINERAL SPRINGS IN FRANCE

AUTHORISATION

A MINERAL spring cannot be worked medically and com-
mercially without a special *authorisation* from the Minister of
Commerce and Public Works. Even this authorisation is
only granted after a favorable opinion has been pronounced
by the Academy of Medicine.

Any landowner who discovers a mineral or medicinal spring
upon his ground is bound to inform the Academy of Medicine
of the fact, in order that an inquiry may be made; and according to the report of the Commissioners appointed by that
body, the distribution of the water is permitted or prohibited.

The Academy of Medicine directs the examination of the
various documents, and the analysis of the water, to be carried on in its laboratory. The report presented to the Academy states the composition of the water analysed, and con-
cludes, according to its composition, whether or not it is
advisable to authorise its use for medical purposes. The
ministerial decision is given in conformity with the conclusions presented by the Academy of Medicine.

Mineral springs may be declared to be *of public utility* by
an official decree, issued by the Council of State; but if a

mineral spring once declared *of public utility* is afterwards
worked in a manner to endanger its preservation, or if the
mode of working it does not fully satisfy all sanitary requirements; the Council of State can order that its use should be
suspended or altogether discontinued, or even that the spring
itself should be destroyed.

Official regulations fix the forms and conditions of the decla-
ration *of public utility*, of the constitution, authorisation, and
inspection of the springs of natural mineral waters and the
Establishments in connection with them, and of the general
conditions of order and salubrity which all public establishments must fulfil.

MEDICAL INSPECTION AND CONTROL OF THE SPRINGS AND ESTABLISHMENTS

A Medical Inspector is appointed to every locality possessing one or more Establishments of natural mineral waters,
the employment of which is considered to require special
superintendence.

During the season for taking the waters, the Inspector
exercises a constant surveillance over all the parts of the
Establishment appropriated to the use of the waters and the
treatment of patients, as well as over the carrying out of all
the regulations referring to these matters. Such regulations, however, are not in any way to restrain the liberty of
patients to follow the prescriptions of their own physicians,
or to be accompanied by the latter if they desire it.

Medical Inspectors must have no pecuniary interest in any
of the Establishments of which they have the supervision.

The Medical Inspector and the Government Mining Engineer are bound to inform the Prefect of any disobedience to or
infringement of the regulations, which may come to their
knowledge. They propose, each in his own department, whatever measures may appear to them to be necessary.

One month before the commencement of each season, the
proprietors or managers of the mineral-water Establishments
send to the Prefects a detailed tariff of prices, according to

the different modes in which the waters are administered, and the use of relative accessories.

No change in this tariff can be made during the season.

Under no pretext is any price higher than that named in the tariff to be demanded, nor is any charge not mentioned in the tariff to be made for the use of the waters.

This tariff is always kept posted at the principal entrance and in the interior of every Establishment.

At the close of the season the proprietor or manager of every Establishment of mineral waters forwards to the Medical Inspector a statement of the number of persons who have availed themselves of the Establishment. This statement is sent, with the remarks of the Medical Inspector, to the Minister of Commerce and Public Works.

The proprietors and managers are bound to give to all functionaries delegated by the Minister free access to the Establishments and springs, and to furnish them with all the information necessary to the accomplishment of the special mission entrusted to them.

The Mineral Springs of France offer, therefore, to the public a double guarantee :

1st. The official analysis of the waters, which is obligatory before their use is authorised.

2nd. The constant supervision of Medical Inspectors appointed by Government.

ITINERARIES

FROM PARIS TO THE PRINCIPAL FRENCH THERMAL STATIONS.

NOTICE.—*The following Itineraries have been compiled with the greatest care; but as changes frequently take place in the various services, travellers are advised to consult the Time Tables issued by the Railway Companies.*

CENTRAL DIVISION

Royat.—(265 miles.)   Lyons Terminus (Gare de Lyon). Railway to Royat direct, or to Clermont-Ferrand, where hotel omnibuses for Royat meet all trains.   Fares (about), 1st Class, 41/6 ; 2nd Class, 31/-.

Mont-Dore.—(287 miles.) Lyons Terminus (Gare de Lyon). Railway to Laqueuille or coach from Laqueuille to Mont-Dore. Through tickets, including railway and coach. Fares (about), 1st Class, 50/- ; 2nd Class, 37/6.

La Bourboule.—(294 miles.)   Lyons Terminus (Gare de Lyon). Railway to Laqueuille, and coach from Laqueuille to La Bourboule. Through tickets, including railway and coach. Fares (about), 1st Class, 50/- ; 2nd Class, 37/6.

Saint-Nectaire.—(286 miles.)   Lyons Railway (Gare de Lyon). Railway to Coudes. Fares (about), 1st Class, 44/- ; 2nd Class, 33/3.
From Coudes to Saint-Nectaire. Omnibus meets direct train from Paris. Supplementary service to all trains.

Vic-sur-Cère.—(420 miles.)   Orleans Terminus (Gare d'Orléans). Railway to Vic-sur-Cère direct. Fares (about), 1st Class, 58/6 ; 2nd Class, 44/-.

Châtel-Guyon.—(257 miles.) Lyons Terminus (Gare de Lyon). *Railway to Riom. Fares (about), 1st Class, 40/-; 2nd Class, 30/-.

From Riom to Châtel-Guyon (4 miles) by omnibus or private carriages.

Rouzat.—(257 miles.) Lyons Terminus (Gare de Lyon). Railway to Riom. Fares (about), 1st Class, 40/-; 2nd Class, 30/-.

From Riom to Rouzat (4 miles) by omnibus or private carriages.

Châteauneuf.—(271 miles.) Lyons Terminus (Gare de Lyon). Railway to Riom. Fares (about), 1st Class, 40/-; 2nd Class, 30/-.

From Riom to Châteauneuf (18 miles) by diligence in connection with trains.

Châteldon.—(239 miles.) Lyons Terminus (Gare de Lyon). Railway direct to Ris-Châteldon. Fares (about), 1st Class, 37/6; 2nd Class, 28/6.

From Ris-Châteldon to Châteldon by coach. Fare, 30 centimes.

Vichy.—(227 miles.) Lyons Terminus (Gare de Lyon). Railway to Vichy direct, 9 hours. Fares (about), 1st Class, 36/-; 2nd Class, 27/-.

Sail-sous-Couzan.—(312 miles.) Lyons Terminus (Gare de Lyon). Railway direct to Sail-sous-Couzan. Fares (about), 1st Class, 49/6; 2nd Class, 37/-. Omnibuses from station (2 miles) meet all trains.

Néris.—(207 miles.) Orleans Terminus (Gare d'Orléans). Railway to Chamblet. Fares (about), 1st Class, 32/-; 2nd Class, 24/-.

From Chamblet to Néris (3 miles) by omnibuses meeting all trains. Fare, 75 centimes.

Bourbon-l'Archambault.—(212 miles.) Orleans Terminus (Gare d'Orléans). Railway to Souvigny. Fares (about), 1st Class, 32/-; 2nd Class, 24/-.

From Souvigny to Bourbon-l'Archambault by coach. Fares (about), 1/3 and 1/-.

Saint-Pardoux.—(224 miles.). Orleans Terminus (Gare d'Orléans). Same as Bourbon-l'Archambault, which see.

From Bourbon-l'Archambault to Saint-Pardoux (12 miles) by carriages.

Bourbon-Lancy.—(226 miles.) Lyons Terminus (Gare de Lyon). Railway to Gilly, and coach from Gilly to Bourbon-Lancy (7 miles). Fares, including railway and coach (about), 1st Class, 35/6; 2nd Class, 27/-.

Saint-Honoré.—(192 miles.) Lyons Terminus (Gare de Lyon). Railway to Vandenesse-Saint-Honoré, and coach from Vandenesse to Saint-Honoré (4 miles). Through fares (about), 1st Class, 30/-; 2nd Class, 22/6.

Pougues.—(241 miles.) (Lyons Terminus.) (Gare de Lyon). Railway to Pougues direct. Fares (about), 1st Class, 24/-; 2nd Class, 18/-.

PYRENEAN DIVISION

Eaux-Bonnes.—(446 miles.) Orleans Terminus (Gare d'Orléans). Railway to Pau. Fares (about), 1st Class, 80/6; • 2nd Class, 60/6.

From Pau to Eaux-Bonnes, railway to Laruns-Eaux-Bonnes (within 2 miles of the Establishment),. and carriages.

Eaux-Chaudes.—(446 miles.) Orleans Terminus (Gare d'Orléans). Railway to Pau. Fares (about), 1st Class, 80/6; 2nd Class, 60/6.

From Pau to Eaux-Chaudes by coach or private carriages.

Saint-Christau.—(467 miles.) Orleans Terminus (Gare d'Orléans). Railway to Lacq. Fares (about), 1st Class, 83/-; 2nd Class, 62/6.

From Lacq to Saint-Christau by coach.

Cauterets.—(540 miles.) Orleans Terminus (Gare d'Orléans). Railway to Pierrefitte-Nestalas. Fares (about), 1st Class, 108/6; 2nd Class, 81/6.

From Pierrefitte to Cauterets, one hour, by coach (about), 2/3.

Saint-Sauveur.—(540 miles.)   Orleans Terminus (Gare d'Orléans). Railway to Pierrefitte-Nestalas. Fares (about), 1st Class, 108/6; 2nd Class, 81/6.

From Pierrefitte to Saint-Sauveur, journey 1½ hours, by coach (about), 2/6.

Barèges.—(540 miles.)   Orleans Terminus (Gare d'Orléans). Railway to Pierrefitte-Nestalas.   Fares (about), 1st Class, 108/6; 2nd Class, 81/6.

From Pierrefitte to Barèges by coach, 2 hours (about), 4/-.

Bagnères-de-Bigorre.—(530 miles.)   Orleans Terminus (Gare d'Orléans). Railway direct to Bagnères-de-Bigorre. Fares (about), 1st Class, 88/-; 2nd Class, 63/-.

Capvern.—(724 miles.) Orleans Terminus (Gare d'Orléans). Railway to Capvern direct. Fares (about), 1st Class, 120/-; 2nd Class, 86/-.

Bagnères-de-Luchon.—(737 miles.)   Orleans Terminus (Gare d'Orléans). Railway direct to Bagnères-de-Luchon. Fares (about), 1st Class, 121/-; 2nd Class, 87/-.

Siradan-Ste. Marie.—(750 miles.)   Orleans Terminus (Gare d'Orléans). Railway to Saléchan. Fares (about), 1st Class, 124/-; 2nd Class, 89/-.

From Saléchan to Siradan-Ste. Marie by carriages.

Encausse.—(723 miles.) Orleans Terminus (Gare d'Orléans). Railway to Saint-Gaudens. Fares (about), 1st Class, 119/-; 2nd Class, 85/-.

From Saint-Gaudens to Encausse by coach (about), 2/-.

Aulus.—(528 miles.)   Orleans Terminus (Gare d'Orléans). Railway direct to St. Girons, and coach from St. Girons to Aulus, 18 miles. Through fares (about), 1st Class, 83/-; 2nd Class, 63/6.

Ussat.—(525 miles.)   Orleans Terminus (Gare d'Orléans). Railway direct to Tarascon. Fares (about), 1st Class, 80/-

2nd Class, 60/-. Hotel omnibuses for and from Ussat meet trains.

Ax.—(542 miles.) Orleans Terminus (Gare d'Orléans). Railway to Tarascon, and coach from Tarascon to Ax. 17 miles. Fares (about), 1st Class, 83/-; 2nd Class, 63/-.

Olette.—Orleans Terminus (Gare d'Orléans). Railway to Perpignan (655 miles). Fares (about), 1st Class, 104/-; 2nd Class, 78/-.

From Perpignan to Olette, private carriages and coaches.

Molitg.—Orleans Terminus (Gare d'Orléans). Railway to Perpignan (655 miles). Fares (about), 1st Class, 104/-; 2nd Class, 78/-.

From Perpignan to Molitg, private carriages and coaches.

Le Vernet.—Orleans Terminus (Gare d'Orléans). Railway to Perpignan (655 miles). Fares (about), 1st Class, 104/-; 2nd Class, 78/-.

From Perpignan to Le Vernet, private carriages and coaches.

Amélie-les-Bains.—Orleans Terminus (Gare d'Orléans). Railway to Perpignan (655 miles). Fares (about), 1st Class, 104/-; 2nd Class, 78/-.

From Perpignan to Amélie-les-Bains, private carriages and coaches.

La Preste.—Orleans Terminus (Gare d'Orléans). Railway to Perpignan (655 miles). Fares (about), 1st Class, 104/-; 2nd Class, 78/-.

From Perpignan to La Preste, private carriages and coaches.

Le Boulou.—Orleans Terminus (Gare d'Orléans). Railway to Perpignan (655 miles). Fares (about), 1st Class, 104/-; 2nd Class, 78/-.

From Perpignan to Le Boulou, private carriages and coaches.

SOUTHERN DIVISION

Dax.—(455 miles.) Orleans Terminus (Gare d'Orléans). Railway direct to Dax. Fares (about), 1st Class, 72/-; 2nd Class, 54/-.

Cambo.— (497 miles.) Orleans Terminus (Gare d'Orléans). Railway to Bayonne. Fares (about), 1st Class, 77/-; 2nd Class, 58/-.

From Bayonne to Cambo (11 miles) by coach or omnibus.

Salies-de-Béarn.—(461 miles.) Orleans Terminus (Gare d'Orléans). Railway to Puyoo. Fares (about), 1st Class, 75/-; 2nd Class, 56/-.

From Puyoo to Salies-de-Béarn (4 miles) by omnibus.

Aix.—(535 miles.) Lyons Terminus (Gare de Lyon). Railway direct to Aix. Fares (about), 1st Class, 85/-; 2nd Class, 64/-.

Vals.—(434 miles.) Lyons Terminus (Gare de Lyon). Railway to Aubenas-Vals. Fares (about), 1st Class, 69/-; 2nd Class, 52/-.

Omnibuses and carriages to Vals (1 hour).

Cransac.—(385 miles.) Orleans Terminus (Gare d'Orléans). Railway direct to Cransac. Fares (about), 1st Class, 59/-; 2nd Class, 44/3.

CORSICA

Pietrapola.—(733 miles.) Lyons Terminus (Gare de Lyon). Railway to Marseilles (15½ hours). Fares (about), 1st Class 85/-; 2nd Class, 63/9.

Steamer from Marseilles to Bastia, and coach from Bastia to Pietrapola.

Puzzichello.—(727 miles.) Lyons Terminus (Gare de Lyon). Railway to Marseilles (15½ hours). Fares (about), 1st Class, 85/-; 2nd Class, 63/9.

Steamer from Marseilles to Bastia, and coach from Bastia to Puzzichello.

Guagno.—(698 miles.) Lyons Terminus (Gare de Lyon). Railway to Marseilles (15½ hours). Fares (about), 1st Class, 85/- ; 2nd Class, 63/9.

Steamer from Marseilles to Ajaccio, and coach from Ajaccio to Guagno.

Orezza.—(705 miles.) Lyons Terminus (Gare de Lyon). Railway to Marseilles (15½ hours). Fares (about), 1st Class, 85/- ; 2nd Class, 63/9.

Steamer from Marseilles to Bastia, and coach from Bastia to Orezza.

EASTERN DIVISION

Bourbonne-les-Bains.—(213 miles.) Eastern Terminus (Gare de l'Est). Railway direct to Bourbonne-les-Bains (11½ hours). Fares (about), 1st Class, 35/- ; 2nd Class, 26/-.

Plombières.—(249 miles.) Eastern Terminus (Gare de l'Est. Railway direct to Plombières. Fares (about), 1st Class, 39/7 ; 2nd Class, 29/6.

Luxeuil.—(302 miles.) Eastern Terminus (Gare de l'Est). Railway direct to Luxeuil. Fares (about), 1st Class, 40/- ; 2nd Class, 30/-.

Bussang.—(300 miles.) Eastern Terminus (Gare de l'Est). Railway to Saint-Maurice, and omnibus from Saint-Maurice to Bussang. Fares (including railway and omnibus (about), 1st Class, 45/- ; 2nd Class, 33/6.

Contrexéville.—(232 miles.) Eastern Terminus (Gare de l'Est). Railway direct to Contrexéville (13 hours). Fares (about), 1st Class, 36/- ; 2nd Class, 27/-.

Vittel.—(237 miles.) Eastern Terminus (Gare de l'Est). Railway direct to Vittel (13 hours). Fares (about), 1st Class, 36/6 ; 2nd Class, 27/-.

Martigny-les-Bains. — (248 miles.) Eastern Terminus (Gare de l'Est). Railway direct to Martigny-les-Bains (13 hours). Fares (about), 1st Class, 35/- ; 2nd Class, 26/6.

Bains (*Vosges*).—(284 miles.)   Eastern Terminus (Gare de l'Est).   Railway direct to Bains (10 hours).   Fares (about), 1st Class, 40/-; 2nd Class, 30/-.

Salins (*Jura*).—(250 miles).   Lyons Terminus (Gare de Lyon).   Railway direct to Salins (11 hours).   Fares (about), 1st Class, 39/6; 2nd Class, 30/-.

Uriage.—(400 miles.)   Lyons Terminus (Gare de Lyon).   Railway to Gières-Uriage.   Fares (about), 1st Class, 63/6; 2nd Class, 47/6.   Omnibus to Uriage (4 miles).

Allevard.—(417 miles.)   Lyons Terminus (Gare de Lyon).   Railway to Goncelin, and omnibus from Goncelin to Allevard (6 miles).   Through tickets (about), 1st Class, 67/-; 2nd Class, 50/-.

SAVOY

Aix-les-Bains.—(361 miles).   Lyons Terminus (Gare de Lyon).   Railway direct to Aix-les-Bains (13½ hours).   Fares (about), 1st Class, 57/3; 2nd Class, 43/-.

Challes.—(374 miles.)   Lyons Terminus (Gare de Lyon).   Railway to Chambéry.   Fares (about), 1st Class, 58/9; 2nd Class, 44/-.
From Chambéry to Challes (3½ miles) omnibuses and carriages meet all trains.

Evian-les-Bains.—(326 miles.)   Lyons Terminus (Gare de Lyon).   Railway direct to Evian (12 hours).   Fares (about), 1st Class, 66/-; 2nd Class, 49/6.

NORTHERN DIVISION

Enghien.—(7 miles.)   Northern Terminus (Gare du Nord).   Trains frequently (30 minutes).   Fares (about), 1st Class, 1/2; 2nd Class, -/11.   Return tickets 1/10 and 1/4.

Pierrefonds.—(60 miles.)   Northern Terminus (Gare du Nord).   Railway to Compiègne and omnibus running in connection with trains (thrice daily) from Compiègne to Pierrefonds (8½ miles). Total journey 3 hours. Through fares (about), 1st Class, 9/6; 2nd Class, 7/6.

Forges-les-Eaux.—(72 miles.) Western Terminus (Gare Saint-Lazare). Railway to Forges-les-Eaux direct (3¾ hours). Fares (about), 1st Class, 28/6; 2nd Class, 21/6.

Saint-Amand.—(162½ miles.) Northern Terminus (Gare du Nord). Railway to Saint-Amand direct (5 hours). Fares (about), 1st Class, 25/6; 2nd Class, 19/-.

WESTERN DIVISION

Bagnoles-de-l'Orne.—(140 miles.) Western Terminus (Gare Montparnasse). Railway to Bagnoles direct (6 hours). Fares (about), 1st Class, 24/3; 2nd Class, 18/3.

Château-Gonthier.—(180 miles.) Western Terminus (Gare Montparnasse). Railway to Château-Gonthier direct (6 hours). Fares (about), 1st Class, 28/6; 2nd Class, 21/6.

ALPHABETICAL INDEX OF THE MINERAL WATER STATIONS

WINTERING STATIONS.

PRINTED BY J. E. ADLARD, BARTHOLOMEW CLOSE.

www.ingramcontent.com/pod-product-compliance
Lightning Source LLC
Chambersburg PA
CBHW021805110726
47902CB00006B/1657